THE UNIVERSE IN A BOX

THE UNIVERSE IN A BOX

SIMULATIONS AND THE QUEST TO CODE THE COSMOS

Andrew Pontzen

RIVERHEAD BOOKS   NEW YORK   2023

RIVERHEAD BOOKS
An imprint of Penguin Random House LLC
penguinrandomhouse.com

Simultaneously published in Great Britain by Jonathan Cape,
an imprint of Penguin Random House Ltd, London
First United States edition published by Riverhead Books
Copyright © 2023 by Andrew Pontzen

Library of Congress record available at lccn.loc.gov/2022054457

ISBN 9780593330487 (hardcover)
ISBN 9780593330500 (ebook)

Printed in the United States of America
1st Printing

Book design by Daniel Lagin

To my family

CONTENTS

Does a firm persuasion that a thing is so, make it so?

William Blake,
THE MARRIAGE OF HEAVEN AND HELL

INTRODUCTION

Very occasionally, a single event can expand your mind into a new realm. My astronomer colleagues variously trace their love of space to being given a telescope, spending a night under the stars, or watching the moon landings. The moment that sticks in my memory was discovering, at age seven, my father's ZX Spectrum computer. A musician by profession, Dad worked with a succession of early digital music synthesizers and had a background in electronic engineering. Home computing was the next frontier. The plasticky Spectrum, with its rubber keyboard and rainbow motif, was plugged into an old TV set in the damp basement and soon occupied me for hours a day. It could be instructed to do almost anything, or so it seemed.

The Spectrum stored games and other programs—apps or code, in today's terminology—on audiocassettes. Starting a program was an unreliable process that involved guesswork: fast-forwarding or rewinding to the right point on the tape, typing LOAD, pressing play on the tape deck, and waiting for a few minutes while bizarre sci-fi sounds blared and psychedelic colors flashed on the screen. Eventually the process would come to an abrupt end and, if you were lucky, the game would begin.

One day, somewhere on Dad's numerous cassettes, I found a program called SatOrb.[1] This gem challenged you to launch a satellite around a planet of your choice (you could select any from the solar system). It asked for an initial height and speed, then traced what happened to your hypothetical craft. As a yellow, pixely path slowly drew itself on the black screen, one could start to guess: Will it crash onto the planet's surface, shoot off into space, or achieve the goal of a stable orbit? With practice you could place your vehicle on a suitable trajectory, and it would satisfyingly loop-the-loop—just like the moon, or one of the thousands of artificial satellites that circle Earth.

SatOrb helped kindle my interest in physics and computing, setting me on a course to spend much of my teens in the basement, writing computer code to create programs of my own. I had some books about space, which I enjoyed browsing, and I did glance at the night sky from time to time. But I never thought to ask for a telescope. The blocky, blurry, garish universe inside this little black box seemed to me more real than the far-off, out-there reaches of space.

What I didn't know at the time was that SatOrb is a rudimentary simulation.

Simulations attempt to mimic a real scenario inside a computer, and they are in such common use that they touch every part of our lives. The weather forecasts we all rely upon are based on simulations of Earth's atmosphere; when we drive cars or fly in airplanes, they were simulated and tested before being built; simulations are at the heart of computer-generated special effects for cinema and TV; computer games, architectural modeling, financial planning, and even public-health decision-making are all underpinned by simulations.

My job as a cosmologist involves simulating the entire universe on computers. The goal is to understand what is out there, where it came from, and how it relates to our lives here on Earth. Loosely, we are using the computer in place of a laboratory. Cosmologists can't perform traditional experiments like other scientists might: there is no way to control the universe at large, and even if there were, we would wait cosmic time scales—billions of years—for the results. Simulations offer us a computerized universe where space and time are under our control.

The ability to sculpt virtual worlds is what hooked me on computers, but my life today doesn't involve sitting alone in a darkened room, tapping away at a keyboard. I work with dozens of colleagues here in London and around the world. We publish our results in journals which reach hundreds of others. The whole endeavor rests on the cumulative work of thousands and makes use of powerful computers which fill entire air-conditioned rooms.

There is another difference between my work today and SatOrb: the trajectory of a spacecraft orbiting a planet can be calculated using pen and paper. Manual calculations might be tedious and error-prone, but there is nothing that SatOrb does that can't be accomplished by a determined human being, and no result that SatOrb produces is a surprise to a degreed physicist. It certainly does not reveal any new truths about the reality in which we live. By contrast, when we try to simulate the universe as a whole we really do learn something new, because the results often defy expectations.

I am going to uncover the reasons for that over the course of the book. It is not just about the absurd physical extent of the universe, although that's certainly worth pausing to contemplate. It's hard enough to imagine Earth being almost 13,000 kilometers across, let alone comprehend the size of the sun, into which our planet would fit 1.3 million times over. The sun is just one of hundreds of billions of stars in our home galaxy, the Milky Way, which in turn is one of hundreds of billions of galaxies of various shapes, sizes, and colors, all arranged into a vast pattern known as the *cosmic web*. Simulations reveal how these various structures, despite their inordinate scale, have all played a role in our own origins: as I will show, carbon-based life forms on a small rocky planet couldn't have arisen without these gargantuan support structures. It is boggling. I don't think there is any way truly to come to terms with it.

But the universe isn't just enormous; it is also enormously complex. Simulations are at their most valuable when they trace a kaleidoscope of billions of individual stars, black holes, gas clouds, and dust specks. It can be exceptionally hard to anticipate the collective behavior of such a large number of elements in combination. It does not follow easily from the physics of the individual components.

This stark difference between individual and collective behavior can be appreciated by studying social insects here on Earth. Army ants, for example, swarm to locate colonies of smaller insects, which they then devour. While swarming, they perform extraordinary feats of cooperation, using their bodies to smooth out the terrain, or even to build bridges over uneven ground. And yet no one plots a route to the food, draws up blueprints for a bridge, or dictates where to fill potholes. There is no organizing principle, and yet organized structures still emerge, structures that are hard to anticipate by studying an ant in isolation.

This can be counterintuitive at first, since human social organizations are so heavily based on hierarchies and plans. To human eyes, the collective behavior of the army ants suggests that an executive within the colony formulates strategies to reach prey efficiently. But there is no such individual. There are just lone ants, following simple unchanging rules, like joining an ant bridge if there are many individuals pushing behind and leaving the structure if no others crawl over.[2] The sophistication emerges from the sheer number of individuals following these rules.[3]

Understanding how a coherent, organized universe emerges from a melee of stars, gas, and dust is one of cosmologists' central goals. We build computer simulations based on the laws of nature—gravity, particle physics, light, radiation, and more—in order to obtain predictions that can be tested against night-sky observations. Because their arithmetic is accurate and fast, computers can repeatedly apply simple rules to millions or billions of sub-elements, and reveal for us how a fixed set of rules can give rise to new and surprising collective behaviors.

Simulations help us see the big picture, in which the universe transcends its small-scale laws. By the end of this book, you will have seen just how radical that picture is, describing an intricate cosmic ecosystem upon which our own existence is contingent.

The Craft of Simulations

Setting out to capture the universe inside a computer requires a level of chutzpah. The difficulties are inherent in the goal: understanding how a multitude of tiny influences combine to determine an overall outcome is intrinsically hard. If the simulation misrepresents any one of the influences, even by a small amount, the conclusion might be very wrong indeed. The art of simulation lies in characterizing the individual elements as precisely as possible, while understanding any remaining shortcomings so that the conclusions can be framed with appropriate caution.

These vagaries may come as a surprise. The universe follows a rigid, inarguable set of laws, or so we are taught in school, and it is true in principle that a virtual universe might be constructed by appealing directly to clockwork laws of physics that have been rigorously and extensively verified. This would seem to leave little room for error. The laws are a formalized collection of knowledge and expectations, written in the precise language of mathematics—perfect for translating into computer code. But all is not quite as it seems.

Consider the weather forecast. The presenters who tell you what to expect tomorrow base their expectations on simulations of Earth's atmosphere, combining all the innumerable tiny influences on wind, clouds, and rain to make predictions for the future. But wind, clouds, and rain don't appear directly within the laws of physics, which are instead written in relation to individual atoms or molecules. The weather emerges from the combined effects of the 10^{44} molecules in Earth's atmosphere, and a simulation would seemingly need to know the location and motion of each one.

That is not possible. Any computer's storage capacity is finite, and can be measured in bits, the smallest possible units of storage, corresponding to single switches that can either be on or off. By itself, a single bit is not terribly descriptive but you can store anything if given enough of them. Black-and-white images, for example, can be represented by bits on a grid: switched-on might represent a black dot, and switched-off an empty cell. Numbers, letters, colors, sounds, videos, Facebook friendships: all can be stored as series of bits, and the more bits you have, the more descriptive you can afford to be. A ZX Spectrum had almost 400,000 bits of memory; the laptop on which I am typing has 100 billion bits; some supercomputers have more than 10,000 trillion of them.

This is still nowhere near enough to enable simulations of Earth's atmosphere at the molecular level. If you wanted to store even a single bit of information for each molecule, you'd need a 10^{21} increase in the current storage capacity of the world's computing centers.[4]

So a weather forecast cannot be constructed on the basis of atoms and molecules, and likewise a simulation that tries to tell us about entire galaxies certainly won't be able to track these most fundamental constituents. To fit inside computers, a description of the weather, of a galaxy, or of the entire universe has to lump together vast numbers of molecules, describing how they move en masse, push on each other, transport energy, react to light and radiation, and so on, all without explicit reference to the innumerable individuals within.

If the goal is to mimic reality inside computers, the available resources are laughably inadequate to reach it; the limitations of what can be achieved in practice are often stark. And yet over the last fifty years, as the technology has steadily improved, a growing community of astrophysicists has made cosmological simulation tractable with the aid of crafty physics shortcuts and tricks.

I am going to give you a taste of how these tricks were invented—sometimes through the hard work of lone PhD students who battled to have their ideas recognized; other times by entire laboratories that banded together to crack tough problems; and, in some cases, as a result of national research priorities set by the highest level of governments. Some of the resulting shortcuts are well justified, while others are admittedly more like a stab in the dark. For that reason, not everything within a simulated universe can be taken at face value.

This problem is not unique to cosmology. Humanity has a creeping reliance on simulations, models, and algorithms, with the dividing lines between these categories being fuzzy. I tend to think of algorithms as rules that determine an action to be taken: the way an autopilot corrects the course of an airplane, or a social-media site decides which posts to display, or a satnav calculates which route you should follow, for example. In cases where these decisions aren't totally straightforward, there needs to be an underlying model—a description of relevant phenomena like flight dynamics, human attention spans, or future traffic flow. And if a model involves large numbers of different elements interacting, it is best characterized as a simulation.

A good example of the fine line between algorithm, model, and simulation is financial trading, where inspiration from physics played a major role in the 2008 economic crash.[5] The goal of financial modeling is to predict the future movement of stocks, starting from whatever real-world information is available. Such predictions are impossible in detail; but in the early 2000s, hedge funds fell in love with theoretical physicists and their ability to make informed guesses at the future. Using a few simple assumptions about how individual stocks change value over time, so-called quants built simulations of the long-term market movements which emerge.[6] Based on the resulting predictions, fund managers started placing speculative bets.

But models and simulations are not re-creations of reality, and are therefore only as good as the simplified assumptions on which they rest. When the markets jitter, individual traders panic, trying to second-guess every decision. It is very hard to write rules for how

stocks behave in these circumstances, and bets can turn out spec-
tacularly wrong. Fund managers without the right circumspection,
who were too assured, too blindly committed to the prophecies of the
models or simulations, found their fortunes turned very quickly.

As early as the 1960s, mathematicians argued that the assumptions that underpin financial modeling underpredicted the risks of
rare but catastrophic market falls.[7] Wise financiers of the 2000s
hedged against just such eventualities and took the promises of mod-
elers with a grain of salt. But others were impressed by the sheen of
computer predictions, and lost their investors mind-boggling sums of
money as a result.

The lesson here isn't that simulations are useless but that they are
nuanced, and not to be taken literally. To understand a simulation
requires a deep appreciation for its limitations, and those lie in the
simplifications that separate virtual worlds from the impossibly complicated reality; the better we understand the imperfections, the more
we appreciate what the simulation is really telling us.

In the aftermath of the 2008 financial crash, two leading quants
published a Modeler's Hippocratic Oath: "I will remember that I
didn't make the world, and it doesn't satisfy my equations. . . . I will
not give the people who use my models false comfort about their accuracy. I will make the assumptions and oversights explicit to all who
use them."[8] It's a maxim that should be applied to cosmological simulations, too.

The financial risks of simulating the universe are small compared
with the trillions of dollars staked on stock market bets. Still, cosmologists would like to understand which aspects of our simulations
can be trusted and which cannot. We are trying to construct a story
of creation that is sufficiently accurate to guide wise investment in

new telescopes and laboratories; whatever money is devoted to foundational physics research should be spent shrewdly, maximizing the chance of new discoveries.

The Cosmic Laboratory

There are some fantastical elements in the simulations that I am going to introduce. A good place to start is with dark matter and dark energy: exotic substances, never encountered on Earth, invisible to even the most sensitive telescope, yet seemingly vital to making sense of cosmic history. Without them, simulations are unable to make sense of the universe.

The absurdity of hypothesizing these materials raises the stakes considerably. On the one hand, it increases the onus to show the working of simulations, admit the limitations, and make the case for why we still, on balance, accept the outrageous conclusions. On the other, if one accepts the case for dark matter and dark energy, they are pointing toward entirely new realms of physics, so far untouched by laboratory experiments. There is nothing more exciting to scientists than this kind of frontier; we are driven by the hope that, one day, humanity will know and understand nature's secrets.

Simulations explore the perimeter of contemporary understanding in another respect, related to science's most basic assumption: that everything happens for a reason, through an unbroken chain of cause and effect. From the perspective of a weather forecast, wind, cloud, rain, heat, and cold don't simply appear and disappear; they exist in distinct weather systems that can move thousands of miles before finally dispersing. Accurately charting the weather today is therefore crucial for predicting the weather tomorrow or in a few days' time.

Similarly, the universe doesn't just do whatever it pleases at a given moment, but follows a domino-like progression of events. The chain extends over almost 13.8 billion years, the current estimated age of time itself, but what happened at the start? What toppled the first domino? When building a simulation, we have no choice but to include some informed speculation about what set events in motion.

At least some aspects of the universe's creation are uncontroversial. There is overwhelming evidence that the universe has been expanding throughout its life, and that this expansion has been so extreme that the entirety of space was once microscopic. The expansion can easily be incorporated into simulations, but on its own is not sufficient to define a starting point for them.

Calculations since the 1980s have suggested that any description of our cosmic origins must lie in the theory of quantum mechanics, which is more usually regarded as a description of atomic and subatomic phenomena. Quantum physics has been well tested in laboratories for more than a century, but its implications are highly counterintuitive. The strangest assertion, at the heart of the theory, is that nothing can ever be completely certain. Subatomic particles don't have a precise location within an atom; they jump, seemingly at random, from one place to another.

Since the universe was once so tiny, it has been imprinted with these quantum phenomena. In the early cosmos, matter can't spread evenly because its tendency to jump randomly will create, through sheer luck, some regions with a little more and others with a little less material. According to simulations, these accidental differences act as seeds that grow into every astronomical structure—every galaxy, star, and planet that we can see around us today, 13.8 billion years later.

The upshot is that the universe might easily have looked very different; there is a strong element of chance in our own existence, which to my mind is distinctly uncomfortable. Quantum mechanics in our initial conditions dooms any hope of predicting precisely what should be in the sky; simulations can only say what sorts of things, in what sorts of quantities, in what sorts of places, might be present. Yet, from such a weakened starting point, I am going to show how it's still possible to draw surprisingly strong conclusions about the universe.

Depending on your perspective, the expansion of space, the central role for invisible materials, and the influence of quantum mechanics may seem rather unlikely. What makes cosmology particularly difficult is appreciating and accepting the otherness of the cosmos. Reality out there doesn't accord with our human experience, and for good reason: our perspective is limited in scale, in speed, and in circumstance. What would it be like to be microscopic or galactic in extent? How would it feel to travel alongside a light beam? What would happen if we fell into a black hole?

When dealing with all this, it's wise to prepare for some surprises. The materials that sculpt space aren't the ones we know from here on Earth. The rules of time and space that we intuitively understand cease to apply. The distances involved defy comprehension. Even looking through a telescope can be counterintuitive: the light we receive tells us not about the universe today but about the universe in the past. Light travels fast, but still it can take billions of years to cross the vast expanses over which we are peering. Common sense, exquisitely honed on human experience, becomes irrelevant.

The Universe in a Box

To understand the origins of our existence, we need to trace them back into deep space. To fathom deep space, how it nurtures new galaxies, stars, and planets, and how these elements interrelate, we need simulations: mini-universes inside computers. And to build and interpret simulations, we need a meticulous appreciation of physics.

But this isn't physics like it's taught in schools and universities, where there is a menu of compartmentalized topics, a list of equations to memorize, and a correct way to solve every problem. Nobody can simulate every subatomic particle and its influence on every other and so the physics in simulations is, at best, approximate. It is much more messy, much more open to debate, much more *human* than what we teach to undergraduates.

Nor is the physics in simulations very much to do with the future that theoreticians sometimes fantasize about, in which a single equation will come to describe every type of particle and force. Maybe one day we will have such an equation; maybe not. Such a final theory of physics, even if it perfectly describes the behavior of individual microscopic elements of our universe, may have only marginal implications for the overarching narrative of creation. The simulator's quest lies elsewhere, in understanding the way that things—subatomic particles, or stars, or clouds of gas, or whatever—behave en masse. Just as watching a single isolated ant tells you little about the behavior of the colony, so studying abstract equations that describe single particles reveals little about the universe.

Simulations enable a new type of understanding, offloading any hard arithmetic to a computer and allowing humans to focus instead on the connections and relationships that emerge. This, at any rate, is the dream. Getting there requires cosmologists to confront the hidden weaknesses of physics, where there are limits to what we know, restrictions on the computational power at our disposal, and compromises at every turn. Choosing and understanding the compromises is where the excitement and challenge is at its most intense.

The reward is a farsighted vision of our home, the cosmos. And while there is a long way to go before that vision is complete—indeed, it may never be complete—simulations have already taught us about dark matter, dark energy, black holes, galaxies, and the way all these interplay to bring the universe to life. Towering far above their foundations in physics, simulations blend computation, science, and human ingenuity in a way that has transformed what it means to be a cosmologist in the twenty-first century. This is their story.

THE UNIVERSE IN A BOX

1

WEATHER AND CLIMATE

Simulating the entire universe is a tall order, so let's start with something more down to Earth: the weather forecast. In uncertain times, isn't it reassuring to hear an expert predict what will happen tomorrow or next week? Like a scientific soothsayer, meteorologists guide us through life, advising what our day will be like, all with the aid of computer simulations of Earth's atmosphere. More often than most people would credit, the forecasters get it absolutely right.

Forecasters are not so different from astronomers. In antiquity, the professions were inseparable: comets and clouds passing above our heads were both matters for a meteorologist, literally a "studier of the high sky." Later, the work of seventeenth-century physicists made grandiose astronomical phenomena predictable and explainable, and clarified that they had little connection with the weather on our planet. The wind and clouds, despite being so much closer to home, remained stubbornly unfathomable.

The lack of any real progress in meteorology until the twentieth century did not turn astronomers away from its study. Anyone wishing to take precise observations of stars and planets also needs to understand how heat and moisture in Earth's atmosphere subtly bend and distort light. On some nights the stars are reasonably stable, while on others they twinkle, even if there are no clouds. That is not an astronomical phenomenon but a meteorological one, and it is seriously bad news for stargazers: the positions of stars appear distorted and the images of planets are fuzzy. For that reason, astronomers planning their observations will pore over weather forecasts, even if these predictions are now made by meteorological specialists.

Growing up with TV weather reports on tap makes it easy to take forecasting for granted. But despite its humdrum nature, every forecast is a hypothesis and each verified prediction should be regarded as a triumph of science. In 1854, a member of Parliament was ridiculed in the House of Commons for suggesting that weather might eventually be known a day ahead of time.[1] Today, knowing the weather for a week ahead has become routine, and the breakthrough is every bit as important as any other scientific revolution: meteorology touches almost everybody on the planet, is worth billions to the economy, and literally saves lives.[2]

I pointed out previously that it is not feasible to build a simulation of Earth's atmosphere starting from the most fundamental laws of physics, which govern how individual atoms and molecules operate. Simulations need instead to describe a higher-level view of how gases move, heat, cool, compress, and expand. Capturing coarse factors like wind or temperature is not too hard, but the problems start when one considers multitudinous details which may have a bearing. For example, think about a tree on a hot day, absorbing the sunlight, sucking water from the ground, and releasing it as vapor into the air. Trees might not initially seem relevant to a forecast but, through the way they soak up light from the sun, alter evaporation, and prevent soil erosion, forests can profoundly change the weather and climate around them.[3] Eight thousand years ago, the region now covered by the Saharan desert had regular monsoon rainfalls; it is possible that human farming removed native vegetation, contributing to turning the whole region into a desert by altering its ability to absorb heat, leading to a runaway effect whereby increasingly dry weather killed the remaining plants.[4] Simulation designers require shortcuts to identifying and including a plethora of surprising effects.

And there is a final ingredient in a successful simulation. Without knowing the weather today, it is impossible to predict the weather tomorrow. On their own, stripped of this information, fancy instructions to the computer are like a purely theoretical set of rules for a board game, dissociated from any specific players or position. Chess grandmasters may know every strategy in the book, but they still can't advise you unless shown the state of the board. What will come next depends on what came before.

This problem of cause and effect turned out to be the main impediment to forecasting before the nineteenth century—just gathering the required information to understand the weather is a large part of the battle. For that reason, simulations' origins lie not with computers but with another electrical invention: the telegraph.

Beginnings

This story opens in the imposing red sandstone Smithsonian Institution headquarters in Washington, DC, now part of the National Mall. The edifice deserves its nickname of the Castle—it is a Gothic masterpiece—but in the late 1850s it stood unfashionably on a partially drained swamp outside the downtown area. Construction had been funded by a bequest to the United States from Englishman James Smithson for the "increase and diffusion of knowledge among men," and the public were baffled and angry. "Nothing has excited more keen opposition," thundered the *New York Times*, "than the costly palace decreed for the accommodation of the Institute. It was unquestionably a gross piece of folly."[5]

Those who overcame their anger and braved the swamp could walk into a cavernous building full of eclectic books, fossils, paintings, and sculptures. But the unique exhibit was a large map of the eastern United States.[6] Each morning at 10 a.m., telegraphic reports were received from weather stations across the country. Then, an assistant would decorate the map using pieces of card—black for rain, green for snow, brown for cloud, and white for fair conditions, giving a snapshot of the current weather. The director, Joseph Henry, would delight visitors by predicting storms in Washington ahead of time, relying on his past experience that weather systems track eastward from Cincinnati.[7]

Across the Atlantic, European navies realized that tracking the weather could give them a strategic advantage. One particularly disastrous storm at the height of the Crimean War in 1854 wrecked at least thirty-seven British and French ships, destroyed encampments, and left the armies' supplies in tatters—"bread, beef, pork, stationery, all rolled into a mass of dirt," according to one eyewitness account.[8] Forewarning of the storm would have been of great value.

Britain, France, and Holland began to invest significantly in weather forecasting.[9] Just as at the Smithsonian, observations from remote parts of the continent would be gathered onto a single map.

In physicists' language, these are known as *initial conditions*: a summary of the current state of affairs, as a starting point for predicting what will happen next. In a cramped office in London, Admiral Robert FitzRoy and a handful of assistants sat among stacks of logbooks and old weather maps. FitzRoy knew from firsthand experience that reliable storm warnings might save lives, so after examining the current observations, he or his assistant transmitted their opinions about the next day to coastal weather stations and newspapers.

The first published weather forecasts appeared in *The Times* of London in 1861. Within months, they were being roundly ridiculed; a typically piqued letter to the newspaper complained that "neither gales nor calms nor direction of wind have been foretold more frequently than a man guessing in a mine might have foretold them."[10] The paper's editors seemed faintly bemused by FitzRoy and his government department's failings: "we, resting upon grave authority, have been made to promise fine weather while the heavens have bestowed upon us a week of fog and deluge."[11]

This doesn't sound so different from our own frustrations over forecast accuracy. We've all taken false reassurance from a forecast, heading out without an umbrella, only to get drenched later in the day. But in truth, these early attempts were dire. American meteorologist Cleveland Abbe estimated in 1869 that only 30 percent of the European predictions for a day ahead were accurate—and yet regarded that as a great encouragement to start a formal forecasting service for the United States, extending the Smithsonian's efforts across the continent.[12]

Abbe was confident that FitzRoy's problems lay with insufficient knowledge of the initial conditions. In England, storms barreled in from the Atlantic, unseen by any weather station until too late. By contrast, inland parts of the United States had significant warning of the approaching weather from a network of stations across the country: "We shall be able to assert with confidence the nature of the weather for one, two or four days in advance," he wrote.

Laws of Nature

The initial conditions are vital, but only tell you so much, and Abbe realized there was much more to it. Originally an astronomer, he was perpetually thirsty for knowledge; a friend reported that he rose early each morning to read consecutively through *Encyclopedia Britannica*, which ran to more than twenty volumes.[13] Whether he completed the task is lost to history. Certainly he loved to talk about philosophy, art, and literature, but found himself increasingly preoccupied with meteorology, and settled on it as his profession.

By 1901, he had gathered what he thought should form the basis of a truly rigorous weather forecast.[14] He observed that his own forecasts to date "merely represent the direct teachings of experience; they are generalizations based upon observations but into which physical theories have as yet entered in only a superficial manner if at all. They are, therefore, quite elementary in character as compared with the predictions published by astronomers."

Abbe's proposal for forecasting was to discard the useful but imperfect folklore of how storms travel, and instead to trace through the consequences of a small number of physical principles. Although the first digital computers were almost half a century away, he came close to describing a simulation. At the core of Abbe's scheme were three equations of fluid dynamics; these weren't new (they are called the *Navier–Stokes equations* after two nineteenth-century scientists) but it was the first time anyone suggested applying them systematically to weather prediction.

While the word "fluid" evokes a liquid like oil, gasoline, or water, to a physicist nearly everything is a fluid—air, glacier ice, plasma in the sun, and the gas in galaxies are all examples. Consequently the three Navier–Stokes equations govern the behavior of materials which would seem at first to have little in common. They are even known as *laws of fluid dynamics*; while they don't refer to the most fundamental particles of nature, they nonetheless earn their status as law because their implications are so universal. The first law states that fluids cannot appear or disappear. When it comes to weather, air is the fluid; it's invisible to our eyes but is certainly there, 25 trillion trillion molecules in each cubic meter. The vast majority of molecules around you now will remain somewhere in the atmosphere indefinitely.*

The idea of conserving material captures something vital: weather primarily consists of shunting material from one part of the world to another. This is at the heart of the early weather forecasting that tracked storms across continents, but it is also a universally powerful insight. On a cosmic scale, winds can blow for billions of years and pile material like a gargantuan snowdrift. Where the winds converge, a galaxy is built; where they diverge, a giant, destitute, empty space remains—a cosmic void. I am going to say much more about these voids later, but for now it is worth acknowledging that there are gaping holes in our universe. Thanks to the conservation law, you can

---

* In reality, a small fraction of the gases in the atmosphere can be lost into space, absorbed into growing plants, deposited into the ground, and so on. The law is not absolute, but it is reliable enough to be exceptionally powerful.

already see that their existence is a necessary counterpart to the billions of bustling galaxies that enable light and life.

Like all the best ideas, the conservation law is simple and yet powerful. But it isn't the whole story for weather or any other simulation. The second Navier–Stokes equation expresses how different parts of a material push upon each other—in other words, it is all about forces. Weather forecasters often talk about pressure, which is another name for jostling at a microscopic level. On the scale of entire weather systems, high pressure pushes material outwards, while low pressure tries to suck it in. For an accurate prediction, a simulation will have to take into account other forces, too, including gravity and the centrifugal and Coriolis effects which are associated with Earth's spin. The combined effects on the weather are far from straightforward.

To appreciate just how strangely forces in fluids can behave, take a sheet of paper and place it on the table in front of you. Take hold of the closest two corners. Raise the sheet, keeping it flat and parallel to the table, then let the far edge droop downward, as it naturally tends to. Bring the close edge directly under your lips, blow vigorously, and watch the floppy sheet unfurl upward, re-straightening itself. It's remarkable—it seems reasonable that blowing *under* the paper would straighten it out, but blowing *over* it?

When air flows over a curved surface like a piece of paper, the wing of an airplane, or the surface of Earth, it generates forces pushing in surprising directions. The converse is also true: the wind often doesn't flow in the same direction as one might guess. Air does not rush straight from high to low pressure regions, but is bent into circular flows by the spin of Earth itself, so when you see low pressure on the TV weather, you know the wind will be spiraling around its core. That allows storms to be much more destructive and long-lasting than they would otherwise be. Without the spin of Earth, air would flow straight toward low pressure, and storms would blow themselves out almost as soon as they formed.

We can count ourselves lucky that hurricanes aren't worse. Jupiter hosts the Great Red Spot, a single storm which is around the size of Earth, and which has been blowing for at least 200 years. On an even grander scale, solar system planets have been orbiting the sun for billions of years; the gravitational force continually pulls the planets inward, as though to the eye of a storm, but only succeeds in bending their path into circles. Forces create curved motions which must be taken into account meticulously if a simulation of weather, the universe, or of anything much else is to be accurate.

All this motion requires energy, and that is the third consideration for fluids. Whether it's on Jupiter or Earth, the vast majority of energy in our solar system comes from sunlight, and without it storms and hurricanes would not occur. On the other hand, the sun is also essential to our survival. If our star were to be extinguished tomorrow, Earth would quickly cool, becoming uninhabitable after a week or two, with temperatures eventually plummeting to around -400°F.[15]

Energy is sometimes a help and other times a hindrance for the development of the cosmos, just as it is for life in the solar system. The faint glimmers of light reaching from stars into the remotest nooks of the universe are enough to heat the tenuous gas that's out there. More destructively, supernova explosions rip holes light-years across. Even black holes have to be included in the grand cosmic energy balance that controls the lives of galaxies. So all three laws of which Abbe was aware—about conserving materials, about calculating forces, and about keeping track of the constructive and destructive effects of energy—are important just as much in the farthest parts of space as they are here on Earth.

Solving Equations

It was one thing for Abbe to recognize that the three Navier–Stokes equations should lie at the foundation of weather forecasting, but another to use them in practice. The equations themselves are succinct and elegant: the principles of conservation, force, and energy can be written in a beautifully compact set of symbols. I still have the undergraduate notebook in which I first scribbled them down from the board in three innocuous-looking lines. But solving them is another matter.

At some point in our lives, we were pretty much all taught to solve equations with a single unknown x. In the more advanced classes, we learned to solve equations with more than one quantity—simultaneous equations in, say, x and y, where each letter stands in for a number that is to be found. But the Navier–Stokes equations don't just have two or three unknown quantities—they are differential equations, which can have an infinity of unknowns.

To see why, imagine waves crashing onto a beach, a scenario that can be described by Navier–Stokes. One of the symbols that appears in the equations represents the speed of motion. But there isn't a single number for that. Water isn't moving uniformly: every drop can be swelling, crashing, or spraying differently from the others. We still refer to "solving" differential equations, but it's not like solving a more standard equation, because there isn't just a single number to be found for each symbol.

Solutions instead describe patterns of motion that develop, start-ing from a particular scenario (the wave approaching a beach, or the wind in today's weather systems), extrapolating forward in time to predict what happens next. In most scenarios, a full solution would involve an endless list of numbers, one for each part of the immensely complicated motion—in practice, solving the equations in a fully sat-isfactory way is therefore completely out of reach for even the most gifted mathematicians.[16]

This might make it sound like differential equations are a bit im-practicable. But it is possible to find solutions, provided the scenarios are sufficiently simplified, so that the overwhelming detail is excised. Navier–Stokes solutions took up an entire term of my undergraduate degree, pursuing abounding examples: idealized ocean waves, stars, galaxy disks, the atmospheres of exotic faraway planets, and so on. The affable lecturer would meet us in pairs to give feedback on our work, and I was unlucky enough to be paired with the class genius. Reliably, the lecturer would start by telling my partner: "You did it well." Then he turned to me and said, "You"—after which he paused, casting about for some kind words—"you did not."

It wasn't that the three equations were unintelligible: they are per-fectly logical, and their relevance is usually clear. Think about how one might apply the principles to those crashing waves. First, conser-vation: the fact that water can't disappear gives you the characteristic rippling—if the water level is pushed down slightly in one place, it must rise somewhere else nearby. Second, forces: they determine the shape and size of waves, with wind whipping them up and gravity dampening them down. Finally, energy: it is transported from the deep sea into the shallows, causing the waves to crash on the beach.

The tricky part is instead to find the idealizations that tease out simplified aspects of these problems one at a time—the way a steady wind over the sea causes regular ripples to form; or how wind and gravity combine to shape the waves; or why energy is transported differently through deep and shallow water. These kinds of simplified questions can be answered over the course of an hour or two, because the overall motions can be summarized in a handful of numbers.

But I lacked the patience, and wasn't sure it was worth persisting. The results can only be taken as a broad-brush picture of how nature might behave; when you make such sweeping approximations, the results are indicative at best. Only a shadow of the magnificent complexity of the cosmos can be captured and, before computers, that was the limit of humanity's ability to turn the abstract laws of fluid dynamics into concrete insight. I realize now that I should have tried harder. It's not just an annoying exercise for undergraduates; professional scientists boil problems down to their essence in this way, and it can be enlightening, even if it usually isn't particularly accurate.

Still, Abbe wasn't in search of abstract enlightenment but a practical ability to predict the weather starting from the fluid equations. He realized that breaking the problem into simpler, idealized scenarios wouldn't do. And yet, he was still confident that scientific weather forecasting could and should be attempted—a fierce optimism was at once his strength and weakness. "He seldom took into consideration such obstructive factors as lack of time or want of opportunity," one of his obituaries noted; many of his projects were impossibly ambitious.[17]

In the specific case of forecasting, Abbe's optimism would turn out to be well placed, as other leading meteorologists around the world started to adopt the vision. Less than twenty years after Abbe's 1901 paper, the Scottish physicist Lewis Fry Richardson and his wife,

Dorothy, would make the first serious attempt at using the Navier–Stokes equations to forecast weather.

Simulations Without Computers

Simulations today take on the mammoth challenge of solving the Navier–Stokes equations without any undergraduate-style simplifications. That is hard enough with computers at our disposal, but the Richardsons used only pen and paper. As though that weren't enough of a challenge, the majority of the calculations were completed by Lewis Fry Richardson while his day job was serving at the front line of the First World War in France, ferrying wounded soldiers to field hospitals.

Raised a Quaker, Lewis Fry Richardson was a staunch pacifist, but resigned from his job in the British Meteorological Office to work for the Friends' Ambulance Unit. Perhaps during his rare rest days, the tedious computations took his mind off matters and helped him feel a connection to home: his wife had been instrumental in collating the all-important initial wind and pressure patterns on a Smithsonian-style grid.[18]

The forecast would take years to complete, and so it wasn't really a forecast in any practical sense: the idea was to show that prediction using the scheme developed by Abbe (and elaborated by another meteorologist, Vilhelm Bjerknes) was possible in principle.[19] Before they were separated by war, the Richardsons collated and tabulated weather reports for 7 a.m. on May 20, 1910. The goal was to use that information to calculate the state of the weather for 1 p.m. on the same day, now long in the past. "Perhaps some day in the dim future it will be possible to advance the computations faster than the weather advances," Richardson wrote, "but that is a dream."

Forecasts of the time were pretty vague, albeit improved over the FitzRoy days. The *Times* weather forecast for May 20, 1910, reads, for the whole of England, "Wind light from some easterly point; changeable, some rain, thunder locally, fair intervals, air rather humid; temperature above the normal." The Richardsons did not aim to do better than these generalities, generating predictions for the average wind, pressure, and moisture in a giant region of 40,000 square kilometers.

That would be fine for proving the point: if the Navier–Stokes equations could compete with the experience of human forecasters, the approach would be worthy of further development. Instead of trying to simplify the equations like an undergraduate is taught to, the Richardsons did the opposite: they unpacked the hidden complexity of the abstract algebra into a set of forms that look like a nightmarish spreadsheet or tax return, crammed with numbers. Written on each form were precise instructions for calculations to carry out—simple operations like adding or multiplying two numbers—alongside stipulations for transferring the resulting numbers onto the next page, where more calculations awaited.

At the end of all this, starting from the grid of conditions at 7 a.m., the calculations would predict the weather at 10 a.m. Using the predictions as input to a fresh set of forms, Richardson pushed the prediction another three hours to the target time, 1 p.m. In simulation language, he was taking two time steps of three hours each.

Modern weather simulations take much shorter steps, measured in seconds rather than hours, to provide the best possible accuracy. Consequently they must also take very many more of them, pushing forward to the days and weeks ahead, increasing the number of calculations far above that which Richardson had to contend with. But the essential principle is the same, even for simulations of the entire universe. We turn the initial conditions into numbers, and transform the three crucial equations into a set of rules for manipulating those numbers. When the rules have been followed, one step has been completed, and the whole process starts again to keep making headway through time.

Computers have the major advantage of being able to calculate quickly and without tiring. Even the processor in your smartphone can perform billions of arithmetical operations every second. Richardson had to calculate everything by hand on occasional days of respite, a few miles behind the front line, using as a desk "a heap of hay in a cold rest billet."[20] There, he could mindlessly follow his own instructions, and he would eventually obtain the very first physics-based weather forecast. It would have been weeks of work even if he had been able to devote himself full-time to the exercise.[21]

But this prototype simulation turned out to be a total failure. It predicted that the air pressure would have risen from 963 to 1,108 millibars in the course of the six hours. There was no need to compare these results with the actual observed weather: the prediction comfortably exceeded the highest ever recorded atmospheric pressure on Earth of 1,084 millibars.[22] Oops.

I can't really imagine what this would feel like in the circumstances, but I sympathize with performing any calculations which lead only to nonsense. (I can still hear that despairing fluid-mechanics lecturer clearly in my mind.) Richardson commented in his book that the result was "spoiled by errors in the initial data for the winds." It sounds desperate but modern analysis suggests it was more or less correct; had his information about the weather at 7 a.m. been more accurate, he could have obtained a sensible lunchtime forecast.[23]

It seems that Lewis Fry Richardson was not at all put off by this experience, as he went on to write a highly technical textbook advocating, not entirely flippantly, that tens of thousands of people be recruited to start computing weather forecasts using the numerical approach.[24] He described an elaborate, purpose-built, giant amphitheater in which forecasters would work, coordinated by a manager in an elevated central pulpit. Kindly, he added that "outside are playing fields, houses, mountains and lakes, for those who compute the weather should breathe of it freely."

In reality, none of that would be necessary. Within the Richardsons' lifetimes, the vision would be realized—not by humans in a grand theater but by electrons whizzing around inside metal boxes.

Computers and Code

I have described two components of a simulation: a set of initial conditions and a rulebook. Starting from the Navier–Stokes equations that describe how air flows around the world, the Richardsons developed a series of arithmetical operations that would step forward in time, incrementally predicting the weather. Because the Navier–Stokes equations encapsulate universal ideas like forces and energy, the same scheme can be used to describe the way that material ebbs and flows throughout our universe.

But none of this is much use without a practical means of performing at speed the vast number of calculations required—something to replace Lewis Fry Richardson's daydream of busy arithmetic masterminds. Humans are expensive, prone to error, and likely to be bored rigid by number crunching. Today's simulations are performed instead by computers which are cheap, reliable, and unable to complain.

The first computer in the modern sense was the Analytical Engine, conceived in the nineteenth century by Charles Babbage. A striking aspect of Babbage's design was that the problem to be solved would be *coded* by a series of holes punched into long strips of card. The patterns of holes would indicate which arithmetical calculations to carry out and in what order, so that, unlike any previous calculating machine in history, this one could be repurposed.

Older machines were sophisticated but inflexible. At the start of the nineteenth century, a surveyor developed a device to calculate land areas; the operator traced the perimeter on a map, and the enclosed area was automatically indicated by a dial.[25] Earlier, in the seventeenth century, mathematician Blaise Pascal developed a machine that could add or subtract any two numbers specified by the user.[26] Even the ancient Greeks developed a machine to predict solar eclipses using a series of meshing gears.[27] Machines like these were ingenious but served only a single designated purpose. Babbage, by contrast, envisioned a machine that could be commanded to perform any desired calculation without changing anything other than pieces of card. The Richardsons could have translated their giant weather forecast forms into coded instructions for Babbage's computer—if only it had actually been built.

Unfortunately, Babbage was not a practical man. His combination of perfectionism and self-righteousness sunk the project: he fell out with the engineer building the machine, constantly revised the designs, resigned in a huff from his professorship in Cambridge, and generally made a nuisance of himself. Babbage had successfully lobbied for public funds to construct his machine, but when no machine was forthcoming, the prime minister summoned him to account for fruitless expenditure.[28] Babbage responded with a diatribe about the failings of government.[29] Inevitably, his purse strings were cut, and the machine was soon forgotten.

Babbage's friend and collaborator Ada Lovelace felt the frustrations of working with him: "I am sorry to come to the conclusion that he is one of the most impracticable, selfish and intemperate persons one can have to do with," she wrote to her mother.[30] Lovelace nonetheless made extensive notes about Babbage's work, and devised examples of how the machine could be instructed to complete specific calculations. She invented the concept of a *loop*, where the same set of instructions would be repeated several times over, each time coming closer to the final result of the calculation, similar to the idea at the heart of the Richardsons' prototype forecast.[31] And she pointed out that Babbage's contraption could shed light on all manner of scientific questions, and even make inroads into art, composing "elaborate and scientific pieces of music of any degree of complexity or extent."[32]

Lovelace actually foresaw something like simulations, writing about how the machine could "facilitate the translation of [scientific] principles into explicit practical forms," and she more broadly anticipated the modern age where computers slip into our intellectual lives as naturally as reading or talking.[33] Pleased with this vision, she jokingly declared herself "very much satisfied with this first child of mine. He is an enormously fine baby and will grow to be a man of the first magnitude and power."[34] The baby was her clear and far-sighted writing, but without a physical machine to embody the vision, it was too abstract to be broadly appreciated.

It would take the best part of a century, but eventually the Babbage–Lovelace vision was independently rediscovered, broadened, and elaborated by Alan Turing. This time, the plans made it past the design stage, largely because progress in electrical engineering made the device more practicable. At every stage of a calculation, Babbage's mechanical computer would have turned heavy metal rods and wheels. By contrast, an electronic computer need only move electrical charges.

On top of the more viable design, military interest played a major role in bringing computing to fruition in Britain and the United States. In 1950, the first machine-computed weather simulation was completed on ENIAC—the Electronic Numerical Integrator and Computer. The machine had been built to aid the US military in the Second World War, and played a key role in the development of nuclear weapons. Its cost ran to $400,000, an enormous sum at the time, but it resulted in an invaluable facility.[35]

It was no coincidence that the warfare-focused ENIAC was also harnessed for weather forecasting. Many scientists, including the nuclear-bomb pioneer John von Neumann, believed that a capacity for accurate weather calculation would eventually lead to new military capabilities. If the natural course of weather events could be predicted, he reasoned, the effects of a human intervention (like spraying aerosols from a plane, or detonating a bomb in the atmosphere) could also be foretold. "One could carry out analyses needed to predict results, intervene on any desired scale, and ultimately achieve rather fantastic effects," he wrote. Controlling the weather could have consequences on unimaginable scales: the connected nature of the atmosphere means an intervention taken by one country might have effects across the entire globe. "All this will merge each nation's affairs with those of every other, more thoroughly than the threat of a nuclear or any other war may already have done."[36]

Such an alarming vision might argue against modifying the environment, but if the Soviets were likely to be developing the capability, the United States had better do so, too: "I shudder to think of the consequences of a prior Russian discovery of a feasible method of weather control," worried one meteorologist in the *Washington Post*.[37] Thankfully, attempts to modify weather for military purposes were only ever trialed on a relatively small scale, were never particularly successful,[38] and were outlawed in 1978 by the United Nations.[39] Even though intentional manipulation of the atmosphere would ultimately be abandoned as too difficult and dangerous, it swayed von Neumann to commit valuable computer time to studying weather.[40] In 1948, he had assembled a team of meteorologists to pursue the first step—twenty-four-hour forecasts.

Like Richardson, the team's first aim was to prove a point rather than to produce anything practical. Also rather like Richardson, the team relied on the women around them: John's wife, Klara Dan von Neumann, is credited as having provided "instruction in the technique of coding . . . and for checking the final code," which is at the heart of the simulation. Code is the instructions to the computer, the punch cards that would have been fed into Babbage's machine, breaking down the problem of solving equations into a series of elementary arithmetic steps.[41]

Code is now everywhere. From the moment you switch on your computer, smartphone, television, internet router, digital camera, car, washing machine, dishwasher, fridge, central heating system, jet plane, air conditioner, space rocket, video recorder, train, CCTV camera, kettle, oil rig, vacuum cleaner, or combine harvester, it's likely to be making use of one or more codes that a team somewhere has written. The internal computers (there can be more than one) may be connected to different bits of hardware—a motor, or a pump, or a display, perhaps—but at their heart they are all pretty similar to vastly miniaturized ENIACs.

But coding for these machines isn't at all like it was in the ENIAC days, when instructing a computer required an understanding of its inner workings. Making even the simplest computations would depend upon knowing exactly the capabilities of a particular machine, so that the coder could express every detail of the way it should calculate. Creating something as complex as a simulation would involve collating thousands of elementary instructions such as add, subtract, multiply, compare, all sequenced and presented in precisely the right way, like building a sandcastle from individual grains. Overall, coding was tedious, error-prone, and increasingly repetitive as technology developed; the slightest change to a simulation, or (worse) switching to a different machine, could entail months of manual copying, adapting, and checking.

Grace Hopper was one of the first to recognize this problem and offer a solution. She had been a mathematician at Vassar College, but quit in the early 1940s to join the Naval Reserve. That landed her in an obscure basement of Harvard University, behind a door under armed guard, working with an early ENIAC competitor known as the Mark I. There, she spent long hours coding solutions to equations specified by navy engineers.[42] It was highly skilled work, but boring and repetitive. Addressing a conference of computer scientists in 1978, she speculated that "most of you would be totally flabbergasted if you were faced with programming a computer, using a Mark I manual."[43]

During the 1950s Hopper and her colleagues hit upon the solution to the tedium of writing code: instead of writing code for a computer, have it write for itself. The initial reaction was derisory, and she had to face down assertions that "it was totally impossible; that all that computers could do was arithmetic, and that it couldn't write programs; that it had none of the imagination and dexterity of a human being."[44]

But she couldn't see why it was so problematic; if a human could provide high-level instructions, provided they were written unambiguously, the computer should be able to take care of translating those instructions into the detailed code that it required for its own operation. It would be more like building sandcastles using a bucket and spade, letting the grains take care of themselves. Computers would now "provide a means that people could use . . . plain, ordinary people, who had problems they wanted to solve," she explained. We "plain, ordinary people," unversed in computer design, would be freed to concentrate on our own specialisms rather than the precise details of how a particular machine works.

And that, today, is what coding is all about: presenting our own specialized information, data, and instructions to the computer, in languages which normally look like an abbreviated, standardized form of English. (Hopper says she terrified the management at Remington Rand, where she worked after the war, by demonstrating that these human-readable computer languages could also be based on French or German. Her team agreed they would stick to English in the future.)[45] There is now a proliferation of different languages that computers can interpret, all with their own idiosyncratic names: Python, Rust, Swift, Java, Go, Scala, C++, among others. The very first that I learned as a child is appropriately named Basic. Whatever the particular dialect, the machine itself figures out the precise way to act upon the instructions, which makes our lives an awful lot easier.

The ENIAC weather forecast, published in 1950, just predated

Hopper's insights, but subsequent forecasts all rely on her new approach to coding, and it is crucial to the later developments that I will describe. Even when expressed in a high-level language, code must still break down the problem into methodical steps which, in the case of simulations, involve stepping forward through time toward a conclusion, just as the Richardsons foresaw. The first ENIAC forecast took eight steps of three hours each to produce the twenty-four-hour prediction. In all, around a quarter of a million individual calculations had to be made, far beyond the reach of humanity before computers.

The results were encouraging, finally providing a prediction which was at least as good as human forecasters could manage. Cleveland Abbe died in 1916, so he never saw his dream achieved, but Lewis Fry Richardson wrote to the project's lead meteorologist "to congratulate you and your collaborators on the remarkable progress" which he and Dorothy agreed was "an enormous scientific advance on the single, and quite wrong, result" that they had earlier obtained.[46]

Resolution and Revolution

The first computerized forecast was impressive to those in the know, but still far from being practical—it took around twenty-four hours to complete. In other words, the machine could only just manage to keep pace with the development of the weather that it predicted.

The room-filling ENIAC performed around 500 individual calculations every second. Within a year, the US Census Bureau was using a computer performing 1,900 per second; within a decade, it was millions, as transistor technology allowed circuits to be miniaturized. Today, tens of billions of these microscopic calculators can be inside a single etched chip, so that the ENIAC forecast can now be completed in a few microseconds on my laptop. But the most powerful computers combine hardware equivalent to tens of thousands of separate laptops into almighty processing megamachines that are no smaller than their ancestors.

Simulations can always put these machines to good use; there is no obvious limit to how much computational power can be soaked up. To see why, imagine a cell phone from twenty years ago; its display looked rough and grainy, because the grid of pixels which made its display was pretty coarse. By making the individual pixels smaller, and having more of them, the images on today's handsets have higher resolution: they are incomparably smoother and more richly detailed.

In the same way, simulations—whether of Earth's atmosphere, or of a distant galaxy—benefit from increased resolution, meaning that they are divided into a larger number of independent blocks. Look at a satellite picture of a storm or of a galaxy and you can see fine detail everywhere. Zoom in and there are details within the details. The more that we can capture and simulate, the more accurate we find our results to be; but that demands more resolution, more independent blocks, and so extra computational power.

Progress means that everyone can now access weather forecasts for a few hours ahead which are localized to within a mile or two, and over the last twenty-five years, the accuracy of the forecast has improved dramatically. Forty years ago, only the one-day forecast meant anything; twenty years ago, forecasts were reasonably reliable out to three days; today, the same accuracy extends five days ahead.[47] These developments can mean the difference between life and death when it comes to forecasting a hurricane.[48]

It is tempting to chalk up the lives saved to the steep rise in computer power and the move to ever-higher resolution, and there is some truth in that connection. But there is more to it. The measurement of the initial conditions by an expanding global fleet of satellites and weather stations also contributes to the improvement and, most of all, the simulations themselves are not just about the rules that I've introduced. There is a hidden dimension to simulations that, for astrophysicists and meteorologists alike, is often the most important part. It is known as the *sub-grid*, and it is here that major strides are being made today.

The sub-grid refers to everything that happens inside single grid squares. Without it, the inner workings of any square are ignored; each is assumed to have completely uniform cloud, wind, temperature, and pressure. The sub-grid attempts to paint detail into that blank canvas, and it's sorely needed. Even if a modern weather forecast divides Earth into grid squares as small as a couple miles across, look up over the course of a hot, humid day and you can see the problem: individual clouds start to form, many of them much smaller than a mile in diameter.

Without the sub-grid, these clouds cannot exist inside the simulation. Either the entire region is cloud-free, or has uniform cloud cover, and nothing in between is permitted. Not only might the resulting forecast miss the possibility of rain but, even worse, it will mischaracterize the sun's heat that reaches the ground and so predict incorrect temperatures. Over time, incorrect temperatures generate incorrect winds and, before you know it, the entire forecast is garbage.

In a world with unlimited computer power, forecasters might try to do without a sub-grid by increasing resolution until the squares are smaller than these bubbling clouds. Future simulations might one day achieve that, but meteorologists will move on to worrying about even smaller-scale processes, all the way down to the microscopic. I already mentioned that the way clouds form over forests depends on the evaporation of water from trees—but that in turn is determined by microscopic pores in leaves which are open and shut by complex biological mechanisms, dependent on the amount of light, the temperature, the water availability in the soil, and so on. Somehow these factors need to be included in the simulation—and they are so tiny compared to any conceivably feasible grid that a sub-grid treatment is the only option. The difficulty is coming up with the right sub-grid rules, ones which will allow the computer to characterize all these detailed considerations in a reasonably accurate but tractable way.

After publication, the Richardsons' work was roundly criticized, not so much for having obtained a plainly wrong answer, but for being unable to capture such small-scale phenomena. One Harvard University professor complained that the work had been doomed to failure from the start because "in our daily weather it is the small-scale phenomenon which works havoc with the forecasts."[49] For good measure, he cautioned that the book was so hard to understand that it would "be quickly placed upon a library shelf and allowed to rest undisturbed by most of those who purchase a copy."

But the Richardsons understood the problem perfectly well and had already put forward the bare bones of the sub-grid solution: bold forecasters will be forced to *invent* rules that replicate the most important missing effects. Since small clouds cannot be represented by the simulation, a new rule is added along the lines of "on a humid, sunny day, after a few hours, some of the sun's rays will stop reaching the ground and rain might start to fall." These sub-grid rules are different in character from the underlying fluid laws; they are specific and narrow, derived from a mix of experience, expectation, and rough calculation rather than purely formal reasoning.

Predicting when it will rain is a job for the sub-grid. One of the ENIAC weather team, Joseph Smagorinsky, pointed out in 1955 that "unlike the normal situation with the other meteorological elements such as pressure, temperature and wind, small-scale precipitation often is of much greater magnitude than the large-scale precipitation."[50] In other words, without knowing every last detail of what's going on in the atmosphere, predicting how much rain will fall is exceptionally hard.

Smagorinsky and his team pressed ahead regardless, developing code that took informed guesses at how much rain would fall averaged over each grid square. Following the now familiar approach of predicting the past, but including rain gauge data in their comparisons, the team checked whether they could get it roughly right—and they could. Smagorinsky was at pains to point out that it wasn't magic: "the fine structure itself is not predicted, but rather the statistical properties of the fine structure," he wrote. Averages come out about right, but the precise patterns of rainfall are unattainable.

The more precise and comprehensive any sub-grid description can be made, the better the simulation will become on average. Water drainage and evaporation, snow and its ability to reflect incoming heat from the sun, melting of the snow, vegetation's multifarious effects, drag on the wind from the rough terrain below; descriptions for all these can be fabricated as a supplement to the original fluid dynamics equations.[51] Today, developing and improving these kinds of schemes is a vast industry, each one being assessed to see how it improves weather forecasts.[52]

Cosmologists are engaged in a similar effort. Simulations of the universe are based on the laws of fluid dynamics, but need to be supplemented by rules of thumb that attempt to paint in missing details. When simulating vast tracts of space, even stars and black holes are relatively tiny and must be included in some approximate way. These details are vital not just for their own sake but for reproducing the big picture, too—a simulation without these ingredients will turn out wrong.

If the dream of a cosmologist is to build simulations that start from a handful of inarguable laws of physics and end with a description of the whole universe, the need for a sub-grid shatters it. The sub-grid is all about creative leaps that fill in the blanks. It is about more than just established laws; it is about informed guesswork, too, which might make simulations appear somewhat scientifically dubious. Understanding what the computers reliably tell us, despite the sub-grid speculations, is part of the artistry of simulation, and something that I am going to talk about in later chapters.

Climate Chaos

There is an even more fundamental limitation on simulations' accuracy. In 1958, when Syukuro Manabe, a young Japanese physicist, was invited by Smagorinsky to join the US Weather Bureau research laboratory, the idea of simulating Earth's atmosphere for more than a day or two ahead seemed to most scientists preposterous. To decide how the weather might look in a hundred years was an even more esoteric exercise, but the powerful John von Neumann had placed Smagorinsky in charge of pursuing exactly that goal. To von Neumann, it was a natural next step in the weather-control research program: he understood that human interventions might be most powerful if sustained over long periods of time, so the entire globe had to be simulated for years ahead. In 1957, shortly after appointing Smagorinsky, von Neumann died (at age fifty-three) from a tumor, perhaps related to his radiation exposure from nuclear weapons development. But Smagorinsky still wanted to understand the climate, and was recruiting the sharpest minds from around the world.

Manabe was the right person to take on forecasting the distant future. "I have very bad driving," he later remarked.[53] "If I start thinking about something, then I'll stop paying any attention to the traffic signals." Soft-spoken and gently self-deprecating, Manabe absorbed himself in the problem for years, while Smagorinsky fought to retain the necessary computers and personnel.

Smagorinsky knew it would be a challenge to achieve convincing results. He had worked with Edward Lorenz, who became famous for asking the question: "Does the flap of a butterfly's wings in Brazil set off a tornado in Texas?"* Playing with simple weather simulations, Lorenz discovered that he obtained profoundly different results for a week or two ahead if he changed his initial conditions by even the tiniest amount—the metaphorical flapping butterfly.

For the few hours ahead, weather is relatively easy to foretell, provided you have a telegraph system. For the days ahead, sophisticated simulations can predict the weather, provided they have accurate initial conditions, high resolution, and a few appropriate sub-grid rules. But over a couple of weeks the smallest disturbance, completely inaccessible to any conceivable measurement or even sub-grid model, can

---

* This is the title of his 1972 lecture to the American Association for the Advancement of Science. Earlier statements made by Lorenz involved seagulls flapping instead, but of course the point is the same.

create a domino-like chain of effects that changes the global weather utterly. This prodigious amplification of initially tiny differences is known as *chaos*.

Chaos dictates that detailed weather forecasts will never extend much beyond a couple of weeks ahead.[54] But climate is not weather. Smagorinsky sensed instinctively that, notwithstanding chaos, the broad features of the long-term climate could be simulated. While a particular storm or heat wave, in a particular place, on a particular day, might be completely unpredictable, trends in the regularity of these events might still be discernible. Later, Klaus Hasselmann, who shared the 2021 Nobel Prize in Physics with Manabe, proved mathematically that these average drifts should indeed be predictable.[55]

By today's standards, Manabe and Smagorinsky's early climate simulations were crude, but they correctly highlighted how an intricate network of different effects—the balance of incoming and outgoing heat, air and ocean circulation, rainfall and evaporation—all combine to determine the long-term future. Manabe recognized he now had "a virtual laboratory of the whole planet": instead of being content to simulate the atmosphere just once, he patiently simulated it over and over again, each time varying different factors that might be important to the stability of our climate. The results revealed how a doubling of carbon dioxide in our atmosphere could increase surface temperatures by 3.6°F, and dangerously change the patterns of weather.

The possibility that carbon dioxide might play a major role in determining climate had first been highlighted by Eunice Foote a hundred years earlier,[56] and had been well known to von Neumann,[57] but prior to the detailed simulations it was hard to be sure. Far from expecting to call attention to a looming crisis for humanity, Manabe initially became absorbed in climate research "just because of my curiosity. I think for many big discoveries, when the research is first started, people would never realize how important their contribution is."

Manabe reached his conclusions in the late 1960s, although the results weren't immediately accepted because other, less sophisticated calculations had produced ambiguous results. After a conference in 1971, a reviewer critiqued that "nearly all the recommendations are for more measurements and more theory," because there were so few points of consensus.[58] It would not be until the late 1970s that the simulations' predictions of global warming became widely accepted, and not until the 2000s that the expectations could be verified by clear real-world data. Today, the reality of climate change is plain, and the problems of consensus lie not with scientists but with legislators.[59] "Understanding climate change is not easy, but it's much, much easier than what happens in current politics," Manabe said after winning the Nobel Prize.

There is another side to climate simulation, one which Manabe was most excited by. Instead of predicting the future, climate simulations can re-create the past. Predicting the future and re-creating the past are not so different—they both involve making assumptions about the amount of carbon dioxide and other gases, then performing the simulation to see what sorts of weather the planet would experience. Using data from ice cores, tree rings, and microfossils it is possible to reconstruct past global atmospheric composition and temperature over thousands or even millions of years, when the state of the Earth was very different, with an average temperature several degrees hotter or cooler. By adjusting the composition of the virtual atmosphere in simulations, Manabe and his team re-created these ancient conditions, showing that historical changes in temperature can be understood in terms of accompanying shifts in the atmosphere's composition.[60] The work on historical climate supplies further evidence that simulations give meaningful answers, despite the huge complexity and the chaos of our atmosphere.[61]

All this provides a first hint as to how cosmologists use simulations. The universe is a messy, chaotic place. We are not in the business of simulating exactly how every detail has unfolded; given how hard it is to predict our own weather, it should be clear that we couldn't capture the entire universe in a computer, even if we wanted to. Still, we hope to reproduce its features in broad strokes, much as climate scientists can sketch past weather patterns.

Just as fossils tell us Earth's history, the light gathered by telescopes is a record of what happened long ago in the universe. Because light travels at a fixed speed, when astronomers look to distant galaxies they are looking back to a time when the universe was a very different place. Our goal for cosmological simulations is to re-create and explain as much as possible in that frozen record. Earth's history is written in the ground, and the galaxy's history is written across the sky.

Earth and the Cosmos

Earth, planet, star, galaxy, the whole universe: whatever it is, the template for a simulation looks pretty similar, which is why I've started with the weather. Simulations step through time from the initial conditions (the weather today, or a cloud of material coalescing to form a solar system, or the aftermath of the Big Bang) to predict how events will unfold. The underlying computer codes are crafted to solve the equations of fluid dynamics describing materials, forces, and energy.

But they also need a sub-grid model, those additional rules that capture all the details that the computer would otherwise miss—raindrops, clouds and soil for meteorologists, or stars, supernova explosions, and black holes for cosmologists.

Weather forecasts have improved markedly over the last thirty years. One of my earliest memories is of a famous storm in October 1987, the worst to hit England in living memory. My family were lucky that just a few tiles were blown from our roof, but eighteen people were killed. Hours before it made landfall, the BBC weatherman Michael Fish announced to viewers that it would be no more than "very windy," an understatement that dogged him for the rest of his career.

"Why weren't we warned?" screamed the *Daily Mail* front-page headline the next day.[62] The paper appeared to blame Fish: "eighty-five-year-old Gwen Hanson, who lives 200 yards from Mr. Fish, had her maisonette roof crushed by a 40 ft elm," it complained.[63] I'll grant that a clearer warning might have helped people prepare, but it seems unlikely that the roof could have been saved. Hanson herself told the paper, "I won't blame Mr. Fish personally," but the paper seemed to imply that she ought to. "Mr. Fish and his family, meanwhile, were out for the day," it added.

A single misstep can land reporters in your street to search for disgruntled neighbors—who would be a weather forecaster? Cosmologists get things wrong all the time, and we rarely get blasted by the *Daily Mail* for doing so. And, as Fish has repeatedly pointed out over the decades since, the storm wasn't missed; it was certainly being monitored over the Atlantic. Unfortunately, its track was imperfectly known and the projections misjudged that it would curve over France, mainly because there weren't enough weather stations in the Atlantic to track its progress accurately. Everyone expected that England would miss the worst of it.

It is an understandable miscalculation, hardly on the scale of the Richardsons' dramatically incorrect first attempts at forecasting, and it wouldn't happen today. There are more weather satellites and stations, even out in the ocean, meaning the initial conditions involve far less guesswork. The computers performing the simulations are more powerful, using increased resolution and improved sub-grid rules to track the small-scale details of heat, wind, and moisture through the atmosphere. And meteorologists' understanding of how chaos leads to uncertain predictions has also progressed. In the 1980s, forecasters had access to a single simulation at a time. Today's forecasters look at dozens of simulations a day, exploring how different storms might track, issuing nuanced guidance and warnings for the week ahead, which cover even those scenarios thought to be unlikely.[64]

Improvements will keep coming, but at some point meteorology will reach a limit, perhaps when we have good forecasts out to ten days or so. Chaos will not permit the indefinite future to be foretold in detail, since we will never know how the wings of all Earth's butterflies are flapping. Similarly, when it comes to the broader universe, it is out of the question ever to have a simulation that perfectly reproduces the night sky in every detail. But climate-science pioneers demonstrated that general patterns can still be anticipated, and for cosmologists that is all the encouragement we need: our goal is to understand an outline history of the universe and its constituents, not to explain individual objects or events.

A simulation of the cosmos is in many ways remarkably similar to one of Earth's atmosphere, but on an unimaginably larger scale.

Weather systems are hundreds of kilometers across and evolve over periods of hours to days; distances and times associated with galaxies are a trillion times greater.

Scale doesn't in itself pose a huge problem. Faced with a blank sheet of paper, one can equally well choose to draw a map of a house, a city, a country, Earth, the solar system, or the Milky Way. The detail you'll get from each is strictly limited—the picture of the Milky Way won't tell you anything about Earth, let alone your house. Similarly, computers can be instructed to focus their efforts on a few days of weather on our single planet, or to spread their attention across billions of stars throughout cosmic history.

But there is a more fundamental difference. The constituents of Earth's atmosphere are 78 percent nitrogen and 21 percent oxygen, with traces of other gases.[65] None of these compounds are particularly abundant in space. In our solar system, the lion's share of the material (around three quarters of it) is hydrogen. And beyond that, in the universe at large, matters get far stranger: it seems that the vast majority of substances out there are unknown to humanity.

In fact, these are the starting points for cosmological simulations: materials which we have never detected in a laboratory. Materials which don't shine, reflect, or cast shadows. Materials which pass through solid rock like a ghost. It sounds a world away from forecasting, but the simulation techniques that I've introduced here will apply in deep space, too. Having seen what can be done to simulate and predict Earth's weather, let's stride confidently into the universe beyond, ready for anything, however bizarre it may at first seem.

2

DARK MATTER, DARK ENERGY, AND THE COSMIC WEB

In 2003, at the start of my third year as a physics undergraduate, I chose to leave laboratories behind and specialize in theoretical astrophysics. It wasn't so much because of a newfound love for astronomy as the agonizing tedium of second-year physics experiments. In my mind now, these were a jumble of lasers, lenses, and little fiddly bits of electronics. Some people seemed to have the knack, and they'd be packing up and heading out of the lab after a couple of hours. I'd still be stuck there as the sun set, poking uselessly at some arcane piece of equipment.

By switching course, I escaped to the astronomy department where one didn't have to do any experiments—nobody can be expected to tinker with the whole universe, after all. But within a few weeks, my small class of defectors came to wonder whether this detachment from reality was dangerous. The majority of our lecturers believed that 95 percent of the universe is made from two hidden ingredients: dark matter and dark energy.

These substances alter the way gravity operates in our universe. Dark matter adds heft to galaxies, changing the way they spin, while dark energy pushes the whole universe apart. In both cases, "dark" is a bit of a misnomer since it incorrectly suggests these materials might obscure light and cast shadows. Instead, they are thought to be transparent—even more elusive than air, they don't shine, reflect, shadow, or have any other direct effect on light.

At least air is easy to trap and study. As a child I had a large plastic cube that held my bath toys, and I loved inverting and submerging it, full of air, then gradually allowing bubbles to escape upward. I could even catch them in another upside-down container as they rose toward the surface, like learning to pour in a parallel, capsized reality. Some experiments are more fun than others.

You can't do anything like that with dark matter or dark energy; no laboratory experiments have yet revealed their existence, and they seemingly can't be held by any container. It may be possible to construct apparatus that provides a gateway into their strange world, but physicists have only the haziest ideas as to how. And so in 2003, my class began to wonder whether defecting from physics to astronomy had been such a good idea. Two invisible, untouchable substances that fix everything we otherwise don't understand—it all seemed a little bit too convenient.

In Rudyard Kipling's *Just So Stories*, strange facts about the real world are given preposterous rationales. Whales feed only on krill because a mariner, annoyed at having been eaten, lodges grates in their mouths. Camels have humps as mystifying retribution from a genie who is troubled by their idleness. Rhinoceroses' skin is baggy because it is filled with crumbs by a cook whose cake they stole. Galaxies spin surprisingly fast because they are crammed with invisible dark matter.

(Kipling didn't write that one, but at first we thought he might as well have.)

Fairy tales are not meant to be taken at face value. By contrast, scientific theories are supposed to say something, if not completely literal, then at least closely related to reality. They are allowed to be based around creative thinking, but must also have some kind of explanatory power that goes beyond magicking away a problem. Ideally they also make a wager: they specify some consequences of the theory which can be tested.

Dark matter and dark energy are indeed serious scientific propositions in this sense. One only starts to appreciate their power by seeing the range of different cosmic phenomena they explain, many of which were predicted before the relevant astronomical observations were made. In the 1980s and 90s, simulations became central to these efforts, while computer-controlled surveys of the sky started cataloging the universe's contents systematically. The agreement between these two strands of investigation is breathtakingly good, provided that invisible dark matter and dark energy are present in the simulations to sculpt the universe in just the right way, but being told this in a lecture is somehow not as convincing as seeing it for oneself. In the 2020s it can be astounding how these theories from the late twentieth century, paired with our modern simulations, continue to make sense of new observations, year after year. That is why they rise far above the status of a *Just So* story.

Over the coming chapters, I will return several times to the accumulation of evidence for dark matter and dark energy. With the aid of simulations, it locks into a coherent set of explanations for seemingly very different phenomena: the sizes and shapes of galaxies; the way that they spin, move, and change over time; the changing expansion rate of the universe; the known facts concerning the opening moments of our cosmos; and the way that all structure today is organized into one giant overarching web. That is a lot to explain in return for two leaps of faith.

Yet frustratingly, the evidence is all still indirect, despite extensive ongoing efforts to find dark matter in the laboratory. It could be a long wait for unambiguous confirmation that these dark substances are as real as the familiar materials around us, and a little historical perspective can help bolster confidence that astronomers haven't succumbed to a kind of collective madness.

Inventing Nature

Precedents for dark matter can be traced back to 1846 and the discovery of the planet Neptune. Too remote to be seen with the naked eye, Neptune did not feature in classical models of the solar system. Even after the invention of telescopes, it had been missed or misidentified in everyone's observations. But in the mid-nineteenth century, several astronomers started to suspect that there may be another planet out there.

The evidence was based on the path of Uranus through the sky. Planets orbit the sun due to its huge gravitational pull, but the path they follow is subtly affected by the existence of other planets. Here on Earth, the influences of the largest—Jupiter and Saturn—change our orbit over periods of tens of thousands of years, explaining the regular occurrence of ice ages.[1]

Using precise astronomical measurements, it became clear in the nineteenth century that Uranus was wandering off course, even once the influence of all other planets was accounted for. In the opening two decades, it was crossing the sky too quickly; by 1822, it was too slow. Two scientists, Urbain Le Verrier and John Couch Adams, independently surmised that the only reasonable explanation was another massive but unseen planet tugging the invisible string of gravity, pulling Uranus back and forth. They both went so far as to calculate the implied location of the missing planet in the sky. Now all that remained was to find it for real using a sufficiently powerful telescope.

Adams, a young student, was shy and ineffectual, and failed to persuade the Cambridge Observatory to conduct a serious search. The deduction of a missing, hitherto unseen planet in the night sky involved exceptionally complex calculations which Adams did not explain particularly well. He did not respond to questions in a letter sent to him by the observatory's director, with the result that nobody was particularly sold on devoting valuable observing time to the search. A half-hearted attempt was eventually carried out once it became clear that Le Verrier was onto the same idea, but it did not result in a discovery.

Le Verrier, despite being forceful to a fault, also failed to persuade his colleagues at the Paris Observatory that a serious search was worthwhile, and it seems possible he had instead alienated and annoyed them. An irascible, self-important figure, whenever faced with difficulties in his deductions, Le Verrier would play his violin feverishly, at any hour of the day or night.[2] His forcefulness would later be instrumental in setting up a French weather forecast service, paralleling FitzRoy's British efforts—but then he sacked all the meteorologists because he couldn't get along with them.[3] As one of his supposed friends remarked: "The Observatory is impossible without him and, with him, even more impossible."[4]

Le Verrier and Adams's prediction of the planet Neptune was

confirmed not in Cambridge or Paris but by astronomers in Berlin on September 24, 1846, after they received a desperate letter from Le Verrier telling them where to point their telescopes. In a sense, Neptune was a prototype of dark matter: something that had never been seen, but which was influencing the remainder of the heavens. With the right telescope pointed in the right direction, it was quickly established that Neptune is as real as any other planet, albeit somewhat farther away and therefore harder to see.

Inventions that fill an inconvenient gap also feature in the history of physics on microscopic scales. In 1930, Wolfgang Pauli, one of the most influential physicists of all time, fabricated a whole new type of particle, the neutrino, without a scrap of direct experimental evidence. In fact, Pauli was not inclined toward doing anything in a laboratory: he was clumsy, and his colleagues even reported that Pauli's mere presence in the room would cause apparatus to break unaccountably.[5] From the safety of his office, Pauli read about experiments and grappled with the foundations of twentieth-century physics, rocking alarmingly back and forth as he cogitated.[6]

Pauli was worried by experiments that showed energy unaccountably disappearing during radioactive decay, in violation of the cherished notion that energy is always conserved. He wrote to colleagues[7] that he had "hit upon a desperate remedy," which was to imagine a new energy-thief particle that had somehow evaded detectors in existing experiments. What came to be known as the *neutrino* was finally found by an elaborate and highly sensitive experiment in 1956, a long twenty-six years between invention and discovery. But that's nothing compared to the Higgs boson—another speculative invention, first theorized in 1964—which was only detected by teams of physicists using the Large Hadron Collider in 2012, forty-eight years later. Patience is a necessary virtue for scientists.

Physics and astronomy advance through the interplay of experiment, observation, calculation, and pure invention. Contriving Neptunes, neutrinos, and Higgs bosons would be meaningless unless followed by careful calculation of their likely effects, and then by experiments and observations that search for these effects. Our understanding of dark matter and dark energy—despite both substances remaining rather mysterious—has also progressed through these stages.

Ingredients for a Galaxy

The case for dark matter starts from appreciating that the entire universe is in continual motion, something which was a source of fascination for astronomer Vera Rubin. Her 1950 master's thesis asked a bold question: Is there any evidence that the whole universe spins? The answer turned out to be "seemingly not": individual objects spin, but her observations showed no evidence for rotation of the universe as a whole. Even asking the question was controversial because cosmologists at the time couldn't conceive how such large-scale spin would be possible even in principle; they regarded the negative answer as self-evident. No academic journal would agree to publish Rubin's observational study.[8]

Later in her career, Rubin became adept at facing down conservative views. She wangled an invite to observe at the cutting-edge Palomar Observatory—where women were technically not permitted. Since the ban on female observers was ostensibly related to the lack of a female toilet, Rubin solved the issue by drawing a picture of a person wearing a skirt, and gluing it to the men's cubicle door.[9] But at the time of her rotating-universe work, she was still a student and decided to cut her losses: "The heated controversy spoiled the fun. People were really very harsh. . . . My way of handling that has just been to go off and do something very different," she recounted.[10]

Rubin remained interested in rotation, but settled on studying how individual galaxies spin, which should have been less contentious. Even the familiar stars, which appear to be arranged in fixed constellations relative to each other, in fact move at phenomenal speeds: they belong, along with the sun, to the Milky Way galaxy, which is spinning at a rate of hundreds of kilometers every second.

Detecting this motion is not easy because, despite the immense speeds, the distances involved are also enormous. Consequently, stars barely appear to move and detecting the spin of entire galaxies requires astronomers to make use of the Doppler shift. Most famous for explaining why the pitch of an emergency vehicle's siren changes as it passes, the Doppler shift also accounts for how the colors of stars change when they are speeding toward or away from us. Given the right technology, the colors of light from a galaxy can be split into a spectrum and measured, and the overall motion of the stars within can then be inferred.

In this way, the fact that galaxies spin had been known since the early parts of the century, mirroring the way that, on a far smaller scale, the planets within the solar system circle the sun. The speed of planetary motion, however, declines steadily with distance from the center. Earth cruises at thirty kilometers per second, while far-out Pluto averages just five. This is a natural consequence of gravity: at larger distances, the force is weaker, and orbiting bodies move correspondingly slower.

The trend with distance should also apply to motions within galaxies. Most of the stars in a typical galaxy lie close to its center; those few that are far in the outskirts should, like Pluto, feel little gravity and therefore move slowly. But through the 1960s and 70s, Rubin studied multiple galaxies, for each one making a spectrum and so measuring the speeds of its own far-flung stars—and it became clear that they zipped along. In fact, such stars move so quickly that they should long ago have been flung out into the empty abyss beyond their galaxy's edge, like a car cornering too fast and flying off the road.

Something must cause this rapid rotation and keep galaxies from falling apart; just as with the mysterious motions of Uranus, the best solution is gravitational tugs from unseen material. But instead of concentrating in a single new planet, the mass has to be spread throughout each galaxy and, especially, far into their outskirts. The extra material came to be known as *dark matter*, and must outweigh regular matter by around five to one if the additional heft is to be sufficient to keep stars on the galactic track.

Rubin was a leading figure in obtaining this 1970s evidence for dark matter, and succeeded in convincing a large fraction of the astronomical community to take the problem seriously. A few people earlier in the century had found tentative signs of missing material and attempted to pinpoint why it might be so hard to see directly. In 1904, famed physicist Lord Kelvin suggested that "many stars, perhaps a great majority of them, may be dark bodies;"[11] in 1930, Swedish astronomer Knut Lundmark speculated that "dead stars, dark clouds, meteors, comets and so on" could add significantly to the mass of galaxies;[12] in the 1930s, the Swiss astronomer Fritz Zwicky talked about early evidence he had discovered for "dunkle Materie" that he thought might be wisps of gas, or cold stars that emit little light.[13]

From a modern perspective, none of these ideas can be right. Largely because of its ubiquity, most cosmologists are instead pretty convinced that dark matter must be made out of some kind of particle—a little like the protons, neutrons, and electrons that comprise the familiar material world, but for one reason or another are very hard to detect directly, echoing Pauli's hard-to-find neutrino. A huge variety of experiments here on Earth are underway to discover what sort of additional particles nature might provide, so far with no success. Understandably, today's undergraduates are just as skeptical as we were in 2003.

Rubin herself was concerned about the lack of progress: "As more and more time passes without direct observational confirmation, I wonder if the explanation is even more complex than we imagine at present," she wrote a few years before her death in 2016.[14] Even the most optimistic cosmologists will admit that, taken in isolation, the unexpected rotation of galaxies is hardly a compelling reason to invent a whole new particle. On the other hand, the 1970s brought cosmological computer simulations into reach, and in the 1980s these simulations would extend the evidence for dark matter from the scale of individual galaxies to the size of the entire universe. It is by understanding these simulations that one can start to appreciate the real power of dark matter as a concept.

Kicks and Drifts

The idea behind simulations is to use physical laws to make a scientific prediction, and so far I have outlined how this works in the case of the weather and climate of our planet. When it comes to cosmological simulations, instead of forecasting how air and moisture cycle through the atmosphere, the simulations are concerned with how stars and other materials move within galaxies and through the universe at large. But how is it possible to ask a computer to predict the behavior of dark matter, when we don't even know what it is? What physical laws can we give the computer as a starting point?

The answer lies in dark matter's defining feature: it is extra material that makes its presence felt through gravity. Luckily, so far as we can tell, gravitation affects all materials in precisely the same way, in contrast with other forces. A fridge magnet, for example, will stick only to certain metal surfaces but gravity is far less picky and, given a chance, will bring absolutely everything clattering to the floor. Whatever dark matter is, we have every reason to believe it will exert and respond to gravitational pulls in the same way as anything else. To make life even easier, it is fair to assume that, like the neutrino, dark matter is nearly oblivious to any other forces—this is the only reason it would behave so differently from regular matter, which coalesces in recognizable atoms and molecules due to electromagnetism. If neutrinos or dark matter were strongly affected by these nongravitational forces, they would be part of the familiar, solid world around us.

Because gravity is a universal force, its effects may be simulated with relatively little attention paid to the type of material under consideration. The first simulation to study gravity's profound influence on galaxies was performed by Erik Holmberg during the Second World War, long before the idea of dark matter was taken seriously.[15] In an echo of the Richardsons' weather forecast, no computers were involved. But nor was it a pure pencil-and-paper affair: leafing through his work makes clear that Holmberg loved technology. He consistently used and even constructed his own complex, ingenious apparatus to tackle problems far ahead of other astronomers—notably, sensitive electronic light-measuring devices known as *photometers* that could scan photographs of galaxies, turning pictures into precise mathematical data for further study. After performing a series of elaborate tests in which he pitted machines against expert astronomers for analyzing galaxy images, he wrote that "the great superiority of the photometer to the human eye is clearly displayed."[16]

Holmberg's simulation came about when he realized that a photometer could be repurposed from a tool for measurement into one of calculation and prediction. Photometers were recent inventions,[17] and not much to look at: seemingly, just a few square centimeters of copper mounted on a wooden stand. But the copper concealed an internal layer of semiconductor—the same type of material that would soon give rise to transistors and the computer revolution. This arrangement generates an electrical current from light, and the strength of the light can then be read via the needle of an electrical meter.

It's far from obvious that this kind of technology could help simulate the role of gravity in our universe, but Holmberg realized that light and gravity are, in a limited sense, interchangeable. The gravitational pull of any mass declines as one moves away and, in a mathematically identical fashion, the intensity of light from any source declines with distance.

Over several weeks in 1941, Holmberg hunkered down in a darkened laboratory and constructed a scale model, a few meters across, of two galaxies. Light bulbs stood in for stars and, by measuring the varying intensity of light, Holmberg determined the resulting gravitational forces. He couldn't replicate the billions of stars inside a real galaxy, but he did use seventy-four bulbs to determine the answer to a single, crucial question: Can gravity can draw together the two galaxies (thirty-seven light bulbs each) and merge them into one? His work was creative, revolutionary, and . . . almost forgotten for thirty years. Some ideas are just too far ahead of the field.

The experiment proceeded in steps, analogous to the way that weather-forecast simulations move through time. Each bulb had a starting position within the experiment and Holmberg also created a written table of their speeds and directions. This didn't mean a literal speed in the experimental apparatus—the bulbs had no means of moving themselves. Rather, the tables recorded the motion in the intergalactic scenario he was trying to re-create, in which two galaxies are fast approaching each other.

To mimic the effects of this movement, he started by manually shifting each bulb along its stipulated direction by the distance it would travel in 1 million years (scaled down to the model size). This is known as a *drift step*, and is still a key part of modern simulations: the stars or other constituents of the simulation drift with their own fixed speed, in their own fixed direction.

That is only half the story because the force of gravity gradually changes the motion of stars. After the drift step, Holmberg therefore settled down to recalculating his tables of motions. This involved measuring the light intensity at the location of each bulb, which in turn told him about the implied strength of gravity from the other bulbs, allowing him to update the figures. This is known as a *kick step*, the idea being that the stars are being booted onto a new trajectory by the gravitational forces. As soon as these first two steps were complete, Holmberg started the whole process again: drift-kick, drift-kick, drift-kick, each cycle of work pushing forward through simulated time by another million years.

In real galaxies, there is no separation into kicks and drifts. Instead, forces gradually change the trajectory of stars so that they follow a curve. The artificial separation replaces curves with a series of straight lines but, provided the size of the steps is small enough, the approximation is excellent. The same idea underlies weather simulation, where the gradual and continuous change of the atmosphere is approximated in the computer by a series of jumps through time.

Given the painstaking work involved—taking measurements, updating tables, precisely relocating seventy-four individual bulbs by hand, then repeating over and over again—it's worth pausing to ask why Holmberg bothered. Even setting aside the design and construction phase, performing the simulation must have been a hugely drawn-out undertaking. Perhaps this was a quiet part of the appeal: in a letter to his fellow astronomer Herbert Rood, Holmberg wrote that he derived great "satisfaction when you found that you could handle everything yourself."[18]

What makes it worthwhile is reaching a conclusion that couldn't be obtained by other means. Without his apparatus, to work out the way that a single star was pulled by the others would have required seventy-three separate detailed calculations. To calculate the pull on each individual star by every other ramps that tally up to several thousand—and he would have had to repeat this formidable undertaking for every one of the several dozen time steps. The task could easily have taken an entire lifetime of calculating; for all practical purposes, it was completely out of reach. The light bulb simulation was still painstaking work, but it gave Holmberg the ability to answer a question that no one else could.

By the end of the simulation, he'd gathered enough data to show that the two galaxies were in the process of gluing themselves together into one, rather than merely flying past each other. He did not

have the resources to continue the experiment further, but he did notice that spiral arms had appeared in his colliding galaxies. These are
the most striking features of galaxies: beautiful, gently sweeping en-
hancements in the intensity of light that look like tendrils of milk in
the galactic coffee. These days we have a huge amount of evidence,
from both observations and simulations, that galaxies do indeed merge,
providing one of several ways to create these spiral structures; Holmberg's conclusion foreshadows our modern understanding.

But it is not just the result. His method, too, presages modern approaches to simulating galaxies and specifically dark matter.

Simulating the Unknown

There are several lessons to be taken from Holmberg's experiment.
First, much like with weather forecasting, there's no strict need for a
digital computer in order to start simulating. Just as importantly,
there's also no need to be too literal about what a simulation represents. If the substitution of bulbs for stars seems natural, it is only
because we can conjure a mental picture of twinkling constellations
inside a pitch-black laboratory.

The bulbs might have looked a bit like stars to a squinting observer, but seventy-four is absurdly short of the typical number of
stars in a galaxy. Each bulb in fact stood in for billions. The trick is
a little similar to political polling: if you want to know who's about
to win an election, you don't have to find out how every individual is
going to vote. Provided you're careful, it's possible to get a good in-
dication from questioning a tiny fraction of the electorate. And so
seventy-four bulbs can represent the gravitational effect of hundreds
of billions of stars.

This detachment from an overly literal mimicking of reality mirrors the way that weather simulations don't track every molecule in our atmosphere, and instead trace the behavior of giant gas parcels. Holmberg's abstraction can be pressed even further: the analogy doesn't rely on whether the mass in galaxies is made from stars or from something else. It captures how material flows through the universe, on the sole assumption that gravity is the most important force in play. Focusing exclusively on gravity would be nonsensical for a simulation of our atmosphere, where pressure and wind are the most vital factors, but it is a superb starting point for understanding space.

And that is where dark matter starts to enter the picture. As the observational evidence mounted that there is at least five times as much dark as visible material, it became natural to reimagine the Holmberg approach as essentially a simulation of dark matter. The fact that dark matter doesn't light up in reality is of no relevance; you could still choose light as a representation of its gravity.

By the time the idea of dark matter was taken seriously in the 1970s, digital computers were sufficiently powerful to take up the baton, so the light trick itself was no longer needed—but the ability to abstract away from detail remained vital. Holmberg's results gave cosmologists some confidence to press ahead with simulations testing whether the gravitational might of dark matter could be responsible for sculpting the whole universe.[19] The building blocks of these simulations are numbers within a digital computer instead of bulbs within a laboratory; the numbers stand in for dark matter as well as stars; and the effects of gravity are calculated through the raw speed of the machine rather than using an analogy with light. But the basic kick-drift technique remains the same to this day, and astrophysicists compare the distribution of material predicted by the simulation with that observed in the real universe, just as Holmberg did.

I have to settle a bit of terminology here. Inside today's digital simulations, there are lots of individual bundles of dark matter that move around, our equivalent of Holmberg's light bulbs. In the 1970s it became common to refer to these as *dark matter particles*, and the name has stuck—but it risks severe confusion. To many physicists, dark matter particle means something else entirely: a physically real particle that they hope, one day, to find using an appropriately sensitive experiment, just as with the neutrino and Higgs boson. Conversely, a particle in a simulation is a stand-in for some material, and no more associated with a real particle than a bulb is associated with a real star. So, I propose a less ambiguous term for the individual chunks of dark matter inside the simulation: *smarticles*, short for simulation-particles.

By the mid-1970s, it was possible to study the behavior of galaxies made from 700 smarticles, requiring around 250,000 gravitational force calculations at each of dozens of kick-drift steps. Computers could deliver results from such simulations within hours.[20] Since then, the largest computers have become more powerful by a stupendous factor in the hundreds of millions. As technology proceeds, it's natural for ambitions to outpace it and so the biggest simulations performed to date have several trillion smarticles. In part this has been driven by one-upmanship, a childish sort of "my simulation has more smarticles than yours," but it is also to an extent driven by real scientific need. Just like a weather forecaster using a finer grid, as we add more smarticles, we are able to start investigating the behavior of galaxies in greater detail.

Adding detail is not the only way to make use of growing computer prowess. An artist can choose whether to paint a detailed portrait, or instead to render a giant landscape. In the same way, astrophysicists can use smarticles to represent a few galaxies with ever-greater precision, or we can choose a larger canvas to start capturing the hundreds of billions of galaxies within the visible universe. All those trillions of smarticles can be used to steadily broaden our horizon until we understand how dark matter (and everything else) spreads through the entirety of space.

Cold Dark Matter

Since the middle of the twentieth century we have known that the universe is around 14 billion years old and that it is expanding, having started as a tiny fraction of its present size. But the expansion does not scatter galaxies at random. Over the 1980s, observations with powerful telescopes showed that galaxies are strung along filaments of a vast "cosmic web" with near-empty voids in between, rather like an enormous cobweb.[21]

The filaments string together dozens or even hundreds of galaxies. Each galaxy is around 10,000 times smaller than the filament itself, so appears as a single bright dot on this scale, and yet that dot contains hundreds of billions of stars, each of which may have multiple planets. So the structure I am talking about is traced through specks of light like dew glistening on a cobweb of dizzyingly gargantuan proportions.

One of the first projects to reveal the bizarre cosmic web structure was led by the astronomer Marc Davis. Comfortable with technology—Davis had funded his college studies by working for a software company—he built an automated, digitalized system for mapping the locations of galaxies in our universe. Not unlike Holmberg decades before, Davis realized that the existing catalogs of galaxies had been assembled haphazardly, and decided to automate the process of scanning the sky with the aid of computers. Inside the telescope dome, "there were wires running all over . . . I didn't do the neatest job in the world, but it did work," he later recalled.[22]

Yet the results seemed a major puzzle: How and why had the galaxies been arranged in this way? Davis turned his attention to finding an explanation, assembling three young researchers to look into the problem using simulations. First was rising star Simon White and his PhD student Carlos Frenk, who had just written a thesis arguing for the existence of dark matter in our own galaxy. Today, on the verge of retirement, Frenk carries an irrepressible, boyish enthusiasm for cosmology. "I can't quite believe it but somehow I've ended up with the best job in the universe," he said in a 2022 lecture.[23]

The team was completed by George Efstathiou, then finalizing his thesis at Durham University and the author of the only computer code in the world that could perform simulations of the necessary scale and sophistication. Efstathiou was in charge of Cambridge's Institute of Astronomy when I arrived to start work on my own thesis in 2005, and to me he was a faintly terrifying figure of authority. But in the 1980s, Efstathiou drove a loud motorbike and wore a leather jacket. The young tearaways became known to the broader community as the "gang of four," a reference to the Chinese Communist Party radicals.[24]

To appreciate one advantage of Efstathiou's code over its predecessors, consider that the universe does not appear to have any edge, as far as we know. When cosmologists talk about the universe expanding, we do not mean there is a bubble of material expanding into an empty abyss; instead, the entirety of space that our telescopes observe is already filled with the cosmic web of galaxies, and yet all of the galaxies are gradually receding from the others. This is very hard to picture mentally, and also a practical conundrum for simulations. How can any finite computer represent a boundless universe?

The solution is to use mathematical tricks to make a small simulated universe look infinite. The closest analogy is the classic arcade game *Asteroids*, in which you pilot a 2D spacecraft around a computer-screen-sized universe, attempting to shoot space rocks before they hit you. If a rock, or your spacecraft, flies through the right-hand edge of the screen, it reappears on the left-hand side, and vice versa. Similarly, flying off the top teleports you to the bottom. Rather beautifully, this provides a gaming universe that has no edge but that is limited in extent and therefore tractable in terms of computational demand. Efstathiou's code implemented this idea, taming the demands of simulating space, placing it inside a miraculous box without walls.

The gang of four combined this universe-in-a-box with the standard kick-drift-through-time approach to simulations and showed how dark matter, with its enormous gravitational influence, would gradually construct a web of material over billions of years. Wherever there is extra dark matter, gravitational attraction sucks in more; conversely, where there's less dark matter, gravity is weak and material will easily be pulled out. This results in a runaway effect: a small patch of dense material will rapidly hoover up everything around it, and over time will form giant structures like galaxies. As these galaxies start to attract each other, some collide and merge, just like Holmberg showed. Others are not close enough to coalesce, but instead line up in a web of galaxies strikingly similar to Davis's maps of the universe.

Like climate scientists, cosmologists can tweak the assumptions within simulations to discover how these different structures respond and whether they match reality. In the 1980s, the burning question centered around neutrinos: Could these mysterious particles account for all the hidden extra mass that the universe seems to harbor? On the face of it, neutrinos were perfect—being completely invisible, abundant throughout the universe, and (unlike any other possible candidate for dark matter) confirmed to exist through laboratory experiments here on Earth.

Those experiments had also showed that neutrinos must be exceptionally light—at most, something like a hundred-millionth of a hydrogen atom.[25] In itself, this doesn't prevent neutrinos acting as dark matter; there are expected to be so many of them in our universe that their total gravitational effect could still be enormous. But the Nobel Prize–winning cosmologist Jim Peebles cautioned that such incredibly lightweight particles move fast; just as a cricket ball is easier to throw at speed than a cannonball, the opening moments of cosmic history knock light neutrinos into a frenzy of movement.[26] Once tweaked to include the resulting rapid motions, the team's simulations confirmed that it was impossible to form the kind of dense, knotty web that had been seen in reality.[27] Neutrinos moved so fast that they shot across the universe instead of creating the required structures.

The discovery was dramatic, because it confirmed that no particle known to physics could account for dark matter: something totally new was required, and it would come to be known, a little cryptically, as "cold" dark matter. The terminology arises from the idea that fast-moving particles like neutrinos are "hot": what we experience as heat is actually rapid motion, albeit usually on microscopic scales.

Conversely, cold dark matter refers to heavy, slow-moving invisible particles. These form structures much more like the real thing. I think of it a little like fondue: if the universe is made of material that's too hot, it becomes thin and runny, but when it's cold dark matter, it lumps everything together into the gloopy webs of structure that telescopes had discovered.

There is a more hidden second conclusion in the simulations: neutrinos must be even lighter than had been suggested in the early 1980s. That is because just having cold dark matter in the simulated universe is not enough; it must be the *dominant* source of gravity. If the neutrinos have too much gravitational heft, they start tugging cold dark matter's web out of shape and the simulations again don't match reality. It's not an option to wish neutrinos away entirely— there are definitely plenty of them out there—so the only viable conclusion is that each individual neutrino must be exceptionally light to minimize its gravitational effects. Today, experiments confirm that the neutrino mass is at least thirty times smaller than physicists had believed at the start of the 1980s: the conclusions from simulations are correct.[28]

These two results catapulted simulations to mainstream attention within the cosmology and particle physics community. Simon White from the gang of four received a rare invitation to travel behind the Iron Curtain to Moscow, where he met Yakov Zel'dovich, a formidable, influential Russian physicist. Zel'dovich had for years been advocating strongly that neutrinos and dark matter were one and the same.[29] Once he saw the simulation results over breakfast in his apartment, he gave a curt nod and changed the topic of conversation.[30] This was, apparently, his way of admitting defeat.

Dark Energy

Cold dark matter goes far beyond a loose story of extra, invisible mass that explains the spin of galaxies; when simulated, it gives rise to a coherent account of how the cosmic web has grown within our universe. Yet the story remains an uncomfortable one—"No physicist worth their salt is going to like this until you can actually discover this dark matter in a laboratory," Marc Davis said in 1988.[31] We are still waiting. The neutrino and Higgs boson searches illustrate that long waits are no surprise but, in the meantime, the universe has started to seem even stranger.

During the 1980s, telescopes around the world continued to improve our understanding of the cosmic web. Davis had been busy focusing on the simulations, but another of the original mapping team's leaders, Margaret Geller, suspected there was more to discover. From childhood, Geller had been fascinated by three-dimensional patterns—she was a regular visitor to her father's crystallography laboratory, which was devoted to inferring the regular latticelike atomic structure of materials. The cosmic web replaced a previously held conviction that galaxies would be scattered at random: "Much of what people seemed to think was known wasn't known at all," she realized.[32] So Geller and two of her colleagues spearheaded a deeper search reaching beyond what Davis had cataloged.

By the end of the decade, Geller had increased the sensitivity of the automated sky survey, finding six times as many galaxies, many of them far fainter and more distant.[33] The extended map revealed that the web of cosmic structure has individual threads stretching over hundreds of millions of light-years. It was yet another surprise to the cosmology community; while the web itself was thought to permeate space, the cold dark matter simulations had shown any individual strand should not be longer than 30 million light-years or so. Clearly something was missing from existing simulations: "A much messier model may be the appropriate one," suggested Geller.[34]

The missing mess turned out to be an antigravity force that draws out ever-longer strands of the cosmic web. The underlying idea has a significant pedigree: Newton and Einstein toyed with repulsive antigravity as a compliment to their respective work on gravitation, but both discarded the possibility as there was no evidence for it.[35] Today, anything pushing the universe apart is referred to as *dark energy*, as a counterpoint to the *dark matter* which pulls galaxies together.

Dark energy's effects must be very weak, since they have no measurable impact on solar system or galactic scales. But even weak dark energy becomes significant on sufficiently large scales. Einstein called it a *cosmological constant*, a gentle but relentless push that generates a gradual yet unstoppable acceleration in the universe's overall expansion.

Some 1980s simulations included dark energy, but only for theoretical completeness: few cosmologists seriously expected it to be a real effect. In 1990, responding to the new generation of galaxy surveys, George Efstathiou pointed out that dark energy's accelerated expansion increased the scale of the cosmic web, as though it had been enlarged in some gigantic photocopier.[36] If a simulated universe were around 80 percent dark energy (the current agreed figure is closer to 70 percent), with the remainder mostly dark matter, the computational and real universes came back into agreement. This implied that if astronomers could measure the expansion of the universe directly, they would find it speeding up in accordance with the large-scale dark energy antigravity.

Eight years later, in 1998, two teams of astronomers announced that they had measured the expansion using the Hubble Space Telescope, and it was indeed accelerating, just as the simulations implied. It was not until I appreciated this point that I finally understood why my undergraduate lecturers in the mid-2000s were so convinced of dark matter and dark energy. It is actually astounding: giant leaps of the imagination stimulated by the combination of theoretical speculation, hard data, and computer simulations, resulting in a prediction which turns out to be accurate.

Prediction might not seem like quite the right word, because all this is about re-creating the past. Predictions in science are normally about the future—somebody who works in particle physics can predict what their experiment will show tomorrow, then find out whether they're right or wrong. For astronomers, predicting the future in this sense is possible but rarely useful. We fully expect that our home galaxy, the Milky Way, will collide with its neighbor Andromeda in less than 5 billion years. It will make for a spectacular night sky. Yet it's not a practical way to gather evidence for or against cosmological theories: nobody wants to wait 150 million generations to see whether they are right.

Such superhuman patience is anyway unnecessary. The universe doesn't change very much over human lifetimes, but what we know of the universe does. For that reason, cosmologists don't typically predict what will *happen* in the future, but what they will *see* in the future. Dark energy kicked in billions of years ago, but until 1998 we didn't know that the expansion of the universe was accelerating; the 1990 simulations made a prediction in this sense.

Since then, telescopes have peered thirty times deeper into the universe, and still the twin explanations of dark matter and energy hold water. When you look out that far, you are staring back in time because the light has taken so phenomenally long to reach you; it is not just today's cosmic web that the simulations predicted but the way it assembled over billions of years. The simulations reproduced something that had already happened, but for which humanity hadn't yet seen the evidence.

Talk about the 1980s and 90s to cosmologists who were there at the time, and they paint a picture of crisis turning to excitement. I was born in 1983, and started studying cosmology in the early 2000s, by which time all these ideas were taken seriously. On one of my first encounters with George Efstathiou in person (or at least one of the first when I dared to speak), I asked whether he believed that all this invisible stuff is out there. He answered with characteristic breezy bluntness: "Yes, of course."

Darkness Visible

I am a convert. If dark matter merely rationalized galaxies' surprisingly fast spin, it would belong in a children's storybook; it's all too easy to concoct a *Just So* explanation for one isolated fact. But galactic rotation is only one of many ways in which dark matter makes its presence felt. Far more significant is the way that dark matter's pull and dark energy's push conspire to produce an overarching structure to our universe, the cosmic web. By experimenting with simulations until the structure looks right, the pioneers correctly inferred significant facts about our universe: that the material we observe directly cannot account for the existence of the web; that neutrinos must be too light to play any significant role; and that the expansion of the universe must be accelerating. This kind of predictive ability is the mark of a successful scientific theory and, while alternative ideas to explain the spin of galaxies have been put forward, no competitor to dark matter has come close to explaining so much.

Today, the picture of a universe with 25 percent dark matter and 70 percent dark energy—just 5 percent left over for the atoms and molecules you, I, planets, stars, and the visible parts of galaxies are composed from—is firmly established through an interlocking network of observations, theory, and simulations, building on all these 1980s and 90s insights. I will present some more of this evidence in the coming chapters.

That is not to say dark matter and dark energy provide final, definitive explanations of what is going on in the universe; they are incomplete in the sense that they don't fully connect to more familiar physics. Neutrinos would have been much more satisfying as an explanation for cosmic structure: we understand not just that they exist but *why* they exist in terms of a broader zoo of subatomic particles which make up our everyday world. Despite their origins as Pauli's desperate remedy, neutrinos are now an integral part of our understanding of nature's constituents, an understanding known as the *standard model* of particle physics.

Dark matter and dark energy are divorced from this standard model, and therefore must be regarded as merely tentative explanations. We don't know how dark matter relates to the familiar world of particles, although there are speculative ideas with exotic names like supersymmetry, axions, and sterile neutrinos (hypothetical cousins of the standard neutrino), each entailing a slightly different version of dark matter. There were high hopes that either the Large

Hadron Collider or a specialized detector might find evidence for supersymmetry, in particular—but in the last few years, with no detections forthcoming, those hopes have started to fade. As it stands, we have few clues as to what dark matter really is, and are completely stumped on what dark energy is all about. Experiments to search for specific new particles continue, but with no guarantee of success anytime soon.

That is frustrating, but it also offers near-limitless opportunities for purveyors of simulations. There are so many possibilities for what dark matter and dark energy might be that people continually code different flavors into their virtual universes, just to see what happens and whether it still agrees with what's truly out there. There is no such thing as perfect agreement; the match between any given simulation and the reality is always a matter of degree. There is always room for improvement, and always the possibility that variations will produce simulations that agree with the universe even better than before.

Can we imagine that dark matter feels, albeit very weakly, some force other than gravity? Or, could it perhaps be moving just a little bit faster than currently believed—not zipping along like a neutrino, but likewise not as sluggish as cold dark matter? Or might dark energy push the universe apart in ever-so-slightly a different way from a cosmological constant?

Following the gang of four's original approach, cosmologists around the world can simulate universes with differing ingredients and compare them with reality; if some change to the underlying assumptions gives a better match to what's really out there, we will be onto something new. At that point, we will be able to issue fresh guidance to laboratories about which sorts of particles they should be searching for.

Some cosmologists would claim we already have seen signs of a coming revolution, with new variants of dark matter or dark energy that match cosmic observations better than before.[37] I'm less sure: comparing a simulation to reality is not easy, and one can easily reach misplaced conclusions.[38] The problem is that late twentieth-century simulations, with few exceptions, focused on the 95 percent of the universe which is dark rather than the 5 percent which is visible. To compare with the reality seen in telescopes required a sweeping assumption: that dark matter's gravity pulls gas and stars in its wake, meaning galaxies always form where dark matter is densest.

It's like painting light onto the dark skeleton of simulations and was, at first, a fair approach. Stars behave very similarly to dark matter from the point of view of a simulation, since the key force—gravity—treats all materials equally; it is reasonable that stars would accumulate where the pull of dark matter is strongest. But this analysis overlooks a difference. A key assumption underlying the simulations and resultant predictions is that dark matter was manufactured a small fraction of a second after the Big Bang. Stars are profoundly different in this respect. They are relative latecomers, first appearing at least 100 million years after the universe began.[39]

Stars, unlike dark matter particles, take time to be fashioned from clouds of hydrogen and helium gas. These gases are subject to forces other than gravity: pressure can push or trap gas clouds, while leaving dark matter free to flow away. Unless a simulation can trace the complex behavior of gas, it cannot predict when and where stars are born, nor where they end up. The idea that stars trace dark matter is a convenient shortcut, but not an accurate one.

By the turn of the twenty-first century, it was clear that the relationship between the invisible and visible components of our universe is complicated. To probe the true nature of dark matter and dark energy, cosmologists had no choice but to first clarify the way that galaxies form and evolve. Simulating 95 percent of the universe seemed impressive but the final 5 percent—the gas, stars, and galaxies—would prove much harder.

3

GALAXIES AND THE SUB-GRID

If you look into the night sky from a city, you'll see only a handful of stars. Venture into the darkness of the countryside, and your eyes slowly adjust to see hundreds, then a few thousand. As your vision adapts further, you will see a faint, light band splitting the sky in two; this is the Milky Way, composed of hundreds of billions of stars that you would need a powerful telescope to pick out individually. And at the right time of the year, on a moonless night your eyes might just make out a smudge of light in the middle of the Andromeda constellation; this is an entire galaxy similar in scale but far beyond our own. Why the universe should be separated into such islands surrounded by near-empty space is a central question for cosmologists.

The Milky Way is our galactic home, and Andromeda its nearest large neighbor, but they are far from the only galaxies. The 1997 film *Contact* starts with a flight over Earth, after which the camera pulls back and our planet starts receding. We pass the moon and Mars, fly through the asteroid belt, and edge past Jupiter and Saturn, until the sun and the solar system are a speck; we glimpse countless stars and clouds of glowing gas, and leave the entire Milky Way floating in a chasm of deep space. *Contact*'s imaginary camera has flown billions of times farther than any real spacecraft has ventured, but the cinematic journey isn't complete yet.

Dozens of new galaxies burst onto the screen; the Milky Way blends into a multitude. Eventually, the screen is filled with dots, each a galaxy beyond our own, some of them smaller, others larger, and each with their own color and shape. *Contact*'s opening reflects a contemporary understanding of the universe: a vast ocean of darkness in which motley bright islands are strung along a cosmic web of structure.

The dark-matter simulations from the previous chapter made sense of the web but could explain little about the galaxies flecked along it. That is because, by definition, a simulation including only dark matter doesn't give us anything to compare directly with what we see through telescopes. Astrophysicists could guess that each sufficiently large clump of dark matter ought to have a galaxy at its center but couldn't begin to explain why galaxies are particular sizes, shapes, or colors. For that, the inclusion of stars and gas in the simulation becomes essential. Adding these ingredients enables a cosmic accounting exercise, testing whether the dark-matter paradigm makes sense when confronted with the range of galaxies observed in reality. More than that, since we live inside a galaxy, the improved simulations are a necessary step in understanding our own history, too. Without knowing how and why gas and stars are strewn through the cosmos, we cannot account for how the solar system and Earth came to be born in the Milky Way.

The existence of galaxies, their histories, their different sizes and shapes: studying all this using computer simulations appealed to me as I started a PhD in 2005. There was something attractive about capturing and studying the building blocks of our universe inside a computer. And it seemed like the right time, too: astrophysicists had just succeeded in simulating galaxies that looked, while not quite like the real thing, at least encouragingly similar.

But as I learned about how these groundbreaking simulations worked, I became disillusioned. Computers are insufficiently powerful for the task at hand; to make even a single galaxy fit inside a computer, essential physics must be simplified into a set of best-guess rules. In particular, the birth, life, and death of stars—the nuclear furnaces that make galaxies visible—are by necessity described loosely rather than in a rigorously principled way.

This is the same idea as the weather sub-grid. Raindrops and leaves are too small and numerous to include in simulations of the Earth, so must be dealt with via some approximation. Similarly, when it comes to galaxies, supercomputers cannot keep track of billions of individual stars inside a galaxy, and the solution is instead to mimic their effects with approximate sub-grid rules. Weather simulations are supposed to serve only a practical purpose, so such shortcuts are fair game. Galaxy simulations, on the other hand, are supposed to tell us about cosmological history, and so the fictional sub-grid seems distinctly more dubious.

The problem will loom large in this chapter. Today, while no longer disillusioned about it, I still spend a lot of time thinking about the strained relationships between reality, physics, and simulation. Computers will never fully capture the richness and detail of our Milky Way, let alone the billions of other galaxies, and so understanding when simulation results should and shouldn't be taken seriously is a skill in itself. Modern galaxy simulations trace how materials assemble into galaxies over time, beginning shortly after the birth of our universe, but their goal cannot be to reproduce every aspect of this long process—such an end is unattainable. Instead, they provide a sketch of cosmological history. The sketch isn't a literal re-creation, but it can be used to interpret the past as seen in reality: our most powerful telescopes peer back through time, because the light from distant objects has taken billions of years to reach us. The faint dots of light received from the ancient universe look very different from the galaxies we see nearby; simulations give us a way to explain why.

To see how astrophysicists have discovered the story of galaxies and to understand what can be trusted in our simulations, it is worth rewinding to the 1960s, when telescopes had seen only a tenth of the way back to the Big Bang. Nobody was paying too much attention to where galaxies had come from, nor how they were changing over time. In fact, it was widely assumed that galaxies had changed little over the past few billion years at least. A lone PhD student, Beatrice Hill Tinsley, would jolt cosmologists out of this complacent belief, charting the path to modern galaxy simulations.

Tinsley's Galaxies

Some works of science are precise, careful, and lucid; others are manifestos and outline a new way of thinking. The text of Tinsley's 1967 PhD thesis somehow manages to combine the two styles.[1] It shows how there is every reason to imagine that galaxies change over time, lays out how to perform simulations that reconstruct and explain these changes, and then concludes that the entirety of 1960s cosmology needed revision.

The preeminent American astronomer Allan Sandage had been using the world's largest telescopes to study galaxies a few billion light-years from Earth with one purpose: to map out the universe's expansion using galaxies' speeds and distances. He was able to measure how fast each moves, and by measuring how bright it appeared in his telescope, he guessed how far away it lay. Any shining object appears brighter when closer, and dimmer when more remote—but to derive a precise distance requires knowing how intrinsically bright a galaxy is, otherwise a bright one far away might be confused with a dim one close by. Sandage worked around this by assuming all his chosen galaxies shone with the same intensity.

Unlike in Holmberg's tightly controlled laboratory where light stood in as a proxy for gravity, Sandage was receiving light from real galaxies; he had no way to verify that galaxies genuinely shine with uniform brightness. In fact, due to the light's traveling time, remote galaxies are seen with a substantial time delay, and were therefore younger when the light left them; Sandage's mapping only made sense if young and old galaxies generate comparable light. He didn't think this was in serious question, and so believed that his measurement of the universe's growth rate would be accurate, extrapolating that cosmic expansion "will cease some 3 billion years hence, after which contraction will begin."[2] The universe, he suggested, would end in 7 billion years with a cataclysmic collision of all the galaxies, stars, and planets.

From Sandage's point of view, to firm up this conclusion required just a few more giant telescopes—but science budgets were being sucked up by the space-flight program. "We're on the verge of rewriting the Book of Genesis," he told the *Wall Street Journal* in 1967. "That's clearly more important philosophically than landing a man on the moon."[3]

While Sandage pleaded his case in biblical terms, Tinsley was undermining his constant-brightness assumption with her careful, eloquent prose. In a letter to her father she pierced to the heart of Sandage's misapprehension: "The calculations also depend on what the various galaxies were like at the time the light left them, the light which now gets to the telescopes, and it isn't necessarily true that they were the same as nearby objects."[4] If galaxies of the ancient universe didn't shine in just the same way as galaxies today, Sandage's projections for the beginning and end of creation were simply wrong.

Tinsley bolstered her case using simulations that she designed, coded, and analyzed. Like all simulations, they started from some initial conditions—just gas, the raw material for forming stars. From this starting point, she instructed the computer to take steps through time, recording how the galaxy changed. But unlike the merging-galaxy simulations of Holmberg, Tinsley was not so interested in how stars move; she was asking instead how the stars within a galaxy are born, live, and die. Holmberg's stars had been constantly shining light bulbs; Tinsley's stars would instead have a life cycle to match the real things. They would even expel nuclear waste products just as actual stars do, adding a variety of elements—such as carbon, oxygen, and iron—to the hydrogen and helium left over from the pristine early universe.

Every star begins as a cloud of gas, drifting inside a galaxy. The cloud is shaped by a delicate interplay between the forces of gravity and pressure, with gravity squashing inward, and pressure pushing outward. Gravity eventually succeeds in making a tightly packed ball, after which nuclear reactions fire up and turn the inert gas into a shining star. After that, the star enters its adult life, but it will still change in color and brightness over time and, once it exhausts its nuclear fuel, may die in a dramatic explosion. The brightest stars live for only a few million years, the blink of an eye in terms of cosmic history.

Tinsley wisely didn't expect her simulation to grapple with any of this directly. Instead of attempting a detailed calculation of how stars are made from individual clouds, she instead instructed the computer that, averaged over an entire galaxy, gas turns into stars at a steady, slow rate to be specified and adjusted by hand. The way that individual stars shine, age, and die she cobbled together from existing pen-and-paper calculations; the computer just needed to add up the effects of all the stars that had formed over the history of each galaxy.

Despite their simplicity, the simulations were sufficiently powerful to prove Sandage wrong: whatever Tinsley tried, there was no way she could produce galaxies that shone with the same brightness over their entire lives. Keeping the galaxies steady would require forming stars at just the right rate to replace those which were dying; that possibility, though, would have implied stars which were a different color to the ones actually seen through telescopes. Tinsley wrote in her 1967 thesis that understanding the origin and fate of the universe "now appears harder than was previously thought, because of the effects of galactic evolution,"[5] the inevitable changes that she had simulated.

The masterstroke in Tinsley's work is that the simulations don't provide a final answer about how galaxies have formed and changed over time—and it doesn't matter that they don't. She wasn't concerned with obtaining a single, correct result; it was clear this wouldn't be possible, given all the complexities involved. Instead, she showed that Sandage's assumption of unchanging galaxies was untenable. A simulation doesn't have to be literally true for it to upend our ideas about the universe.

Friends of Sandage say he was deeply upset; in his mind, an upstart was unfairly attempting to destroy his program.[6] He tried to dismiss the result by claiming real galaxies were incompatible with Tinsley's simulations; in a 1967 lecture in Oxford he asserted that Tinsley's claims were "spurious."[7] But Tinsley knew she was right and responded with a detailed technical paper, showing that Sandage had made an error in his math when comparing to her results.[8] "A significant rate of galactic evolution is not ruled out by any present data," she concluded.

Sandage would only admit that "agreement . . . has not yet been reached," and he continued to cast doubt over Tinsley's simulations and analyses.[9] In spite of that, the work propelled her to international prominence—but then she died of melanoma in 1981, aged just forty. She had published more than a hundred papers which build far beyond her thesis work, defining the entire field of galaxy formation for generations to come. Much of her later work grappled with one fundamental question which still plagues contemporary simulations: How quickly do stars form from clouds of gas?

Without the answer to this question, we don't know exactly how bright each part of a simulated universe should be, or how the brightness should vary over time. As Tinsley's criticism of Sandage's program illustrated, making an erroneous assumption about the brightness of galaxies can also lead physicists to incorrect inferences about the whole universe. Today, the rate at which stars form remains a major source of uncertainty in cosmology, especially when we try to understand the intricate link between visible galaxies and the guiding hand of dark matter.

Galaxies and Dark Matter

The 1980s and 90s were decades of rapid change in cosmology, not least due to the rise of cold dark matter as a compelling explanation for the cosmic web. But remember that the cosmic web was not the starting point for dark matter, which was instead motivated by the need to explain observed anomalies like the over-rapid spin of galaxies. Such observational evidence for dark matter had even resulted in a name for the supposed invisible material surrounding a galaxy: a *dark halo*. Despite sounding oxymoronic, this term accurately describes what astronomers imagine to be present. It may never be possible, but if some future technology allows us to see dark matter directly, we would expect to see a fuzzy auroral haze around every galaxy, extending to around ten times the size of the visible galaxy.

In the meantime, the closest we can come to seeing a dark-matter halo is via *gravitational lensing*, the way light traveling from remote parts of the universe is subtly distorted when it encounters the gravitational influence of a dark halo. This is a long way from seeing the dark halo directly, but measurements of the lensing effect are at least consistent with the presence of an extended haze of material.[10]

The first serious computer simulations of dark matter focused on the cosmic web, which is a much larger-scale structure than an individual galaxy. But as computer power increased, simulations started to show structures of the right scale and mass to play the role of the sought-after dark halos. In the simulations, halos formed where the early universe was at its densest, then grew slowly by gravitationally pulling further material from the nascent web. Often the halos would coalesce, becoming ever larger through a process of continual amalgamation.

It was easy to imagine that gas, too, would be dragged in by dark matter's mighty tug and that it would accumulate until dense enough to form stars. Galaxy mergers, known to occur throughout the universe, would be prompted by the joining of their parent halos. Dark-matter theory was coming full circle: it had been invented based on the observed behavior of galaxies, it had predicted the vast cosmic structures into which those galaxies are arranged, and now it was starting to explain how the galaxies themselves formed and changed over time.[11]

The excitement had to be tempered, though. These simulations weren't really saying anything about visible galaxies, only the dark halos that putatively surround them. They had no stars or gas programmed into them so comparisons with the real universe were based on supposition and guesswork. Simon White and Carlos Frenk—two of the gang of four—decided to address the shortcoming. The task was irresistible because, as White said at a 1988 conference, "Our ideas of how galaxies form are still very uncertain . . . it is not clear that we would recognize a forming galaxy if we saw it."[12] Whether visible galaxies followed the same merging, growing behavior as their dark halos would depend on how exactly gas responded to dark matter's gravity, determining precisely when and where stars formed in cosmic history.

This problem is the same as Tinsley's, but White and Frenk had more rules to invent, because there were entirely new questions thrown up by dark matter. How quickly does gas pour into a dark-matter halo after it has formed? How long does that gas sit around before it is compressed sufficiently to start forming stars? If two halos coalesce, how long does it take before the galaxies within also merge? Bringing together dark matter and galaxy formation was a tall order, and in their 1990 paper, the duo commented that a "disheartening number of ingredients must be assembled to produce a plausibly complete recipe for galaxy formation."[13]

But finding the recipe was becoming urgent, because the Hubble Space Telescope had been launched, and was about to peer deep into the universe in a way never before possible. The observatory, free from the distorting effects of Earth's atmosphere, was able to collect light that had been traveling for most of the age of the universe, and so provide a snapshot of how galaxies had assembled. Theorists knew little about that, having focused instead on how the invisible dark-matter halos had grown. Only if simulations started making clear-cut predictions for the visible parts of galaxies would this gulf be bridged.

Today, when dark matter is almost ubiquitously accepted by professional astronomers, it is hard to imagine what was at stake. The cosmic web evidence for cold dark matter seemed convincing to some specialized cosmologists and particle physicists, but astronomers from the broader discipline were more interested in whether this new paradigm could make sense of the galaxies themselves.[14] If the simulations weren't up to the task, dark-matter cosmology risked being marginalized.[15]

The Hubble Deep Field

In 1995, the Hubble Space Telescope spent ten days over Christmas pointing at a tiny patch of the sky measuring a little less than a tenth of the diameter of the moon. Controversially for such a long exposure, the patch of sky was not known to contain any feature of particular interest: over the preceding year astronomers at the Space Telescope Science Institute had developed a surprising plan to point Hubble at nowhere in particular. But by staring for so long, the telescope could build sensitivity to formerly unknown objects.

The light arriving from dim and distant galaxies comes in a very slow trickle, like sand running through an hourglass. Yet, just as even the slowest-running sand will eventually fill the lower chamber, even the faintest light can be reconstructed into a clear picture if a telescope waits for long enough. When Hubble's ten-day picture was beamed to Earth, it was packed with galaxies.

Hubble Deep Field is the result: imagine a black canvas spattered with bright specks, and a few little swirls painted over the top for good measure. At first sight, the specks might be stars, but in fact each is a distant galaxy. As a twelve-year-old, I saw the image on television and couldn't quite believe there was so much stuff out there, all these thousands of galaxies in that tiny patch. If one were to repeat the exercise covering the whole sky, there'd be 26 million times as many. Yet for the astronomers, the excitement was different: as the British astronomer Richard Ellis put it, what was really striking about the Hubble Deep Field was "the amount of blank sky it contained."[16] A September 1995 newsletter shows the Space Telescope team's own expectations for what they would find, and their mocked-up image was bristling with large, bright galaxies. By comparison, the real galaxies that Hubble uncovered were relatively small.[17]

In one sense, this was a success for cold dark-matter simulations. It was clear that the dark-matter halos around galaxies merge and grow over time; assuming the galaxies within those halos also grow and merge, it follows that galaxies in the distant past were dimmer and smaller than they are now. With cold dark matter in charge of the universe, astronomers should have expected plenty of blank sky even when the world's most powerful space telescope peered across 13 billion light-years.

These expectations hadn't been taken seriously because they were still rather vague. It is one thing to say that galaxies in the past were dimmer and smaller, but another to put numbers on it and set firm expectations for what will be seen through a telescope. Those firm expectations were lacking because the simulations struggled in the face of all the many unknown factors.

One factor above all caused the biggest headaches: how quickly stars form out of gas, the same quandary that left Tinsley guessing in her simulations decades earlier. There is no shortage of gas in the universe, so if gravity were left to its own devices, it would quickly cram dark-matter halos with stars—even small halos would host bright galaxies, the Hubble Deep Field would be packed with light, and today's galaxies would be even brighter still. In the mid-1970s, long before Hubble's striking results, the fact that our Milky Way isn't overflowing with stars caught the attention of Beatrice Tinsley's close collaborator Richard Larson. To Larson it suggested that some reliable means of regulating star formation must be at play throughout the universe. He could identify only one mechanism that could act so ubiquitously, and it's crucial to today's simulations.[18] It is known as *feedback*.

Feedback is the idea that a small number of stars can stop further stars from forming, using a self-defeating loop. Many stars end their lives in a dramatic explosion known as a *supernova*; there is around one explosion per year in the Milky Way, for example. Larson pointed out that the explosion has a side effect of driving gas from the galaxy, removing materials from which new stars would otherwise be born. Think about it like a toilet cistern, in which the rising level of water pushes a valve to stop the flow; once the tank is full, water stops flowing. In a galaxy, once there are enough stars, it can be very hard for more to form.

This concept of self-regulation is powerful but doesn't on its own reveal the exact rate at which stars will form. Feedback's precise effect depends not just on how much gas is available in total but on where it is and how it is moving. In some circumstances, the argument can even be made in the opposite direction: explosions at the right time and place might push together scattered wisps of gas, compressing them into collapsing spheres and so actually *promoting* the formation of further stars. Without a simulation that captures this kind of detail, it's hard to quantify what effect feedback will have.

That is why there were no settled cold dark-matter predictions for what would be seen with Hubble. In the early 1990s, several groups had developed simulations that hybridized Tinsley's approaches with the new concepts of dark halos guiding the accrual of gas and stars in ever-growing galaxies. But with each galaxy represented by just a handful of numbers, these simulations lacked the detail needed to predict the feedback effects unambiguously. It's not that the simulations were incompatible with findings from Hubble: in fact, once the data arrived they were ready to interpret it retrospectively. Just as weather forecasters adjust their sub-grid descriptions of clouds until the weather predictions come out right, it's also possible for galaxy simulators to adjust the sub-grid description of feedback until the numbers of ancient galaxies match reality. By the end of the 1990s, several groups had significant success in this endeavor, finding that very strong regulatory feedback was needed to prevent the simulated Hubble Deep Field from overflowing with galaxies.[19]

It made sense, but it was far from satisfying. What should have been a prediction ended up being something of a retrospective fiddle. Understandably, astronomers were unsure how seriously to take the simulations' explanations for the Hubble Deep Field's emptiness. Richard Ellis worried that "much attention has been given recently to a supposed theoretical triumph in explaining it . . . I think we have to put this result in perspective."[20] He cautioned that playing with feedback until things look right might lead to the correct number of galaxies but for the wrong reasons. His warning was apposite: the challenges that cold dark matter faced were about to get worse.

Grids and Smarticles

When Tinsley envisaged simulating the formation of galaxies, she already understood that it was ambitious in the extreme. The problem was too important to ignore and the difficulty was part of its appeal: in one of her last published papers, she made the problem a positive one: "Essentially every aspect of the subject needs further observational and theoretical study, so galactic evolution will long be a fertile field for research."[21]

By the turn of the century, most simulations of galaxies still ran along the lines that Tinsley had developed—a computerized galaxy consisted of just a few numbers, summarizing how much gas at what temperatures and how many stars of what ages. Dark matter had become part of the mix, but its main effect was to dictate how galaxies are replenished with new gas or merge with their neighbors. That didn't change the heart of the idea. Tinsley's original proposal had stuck, and simulations used a set of best-guess rules to infer how galaxies change over time.

In reality, galaxies cannot be described by a few numbers. It is like talking about storms in terms of the wind speed and amount of rain; it might constitute a helpful summary, but certainly isn't enough to forecast accurately how a particular storm will develop. Similarly, there is no reason why the behavior of galaxies should be predictable without adding substantial detail about how the stars and gas swirl together. This becomes even clearer in light of Larson's theory of feedback: that the vital rate at which new stars form is dictated by the explosive deaths of existing stars. Without knowing exactly where stars and gas lie within the galaxy, feedback might have any conceivable effect, and the astronomy community saw this as a reason to doubt any successes from simulations.

Achieving a better understanding of feedback would entail tracing gas as it drifts and spirals through the universe, just like weather forecasters trace the movement of air and moisture through the atmosphere. One way to include gas in a simulation is to add a vast Richardson-style grid, dividing the universe into cubes. The gas passing through each cube can then be made to obey the three rules of fluid dynamics: conservation, force, and energy. But cosmologists quickly realized this wasn't going to shed much light on the gas within galaxies.[22] The problem is that a grid divides space into equal-sized chunks. That's perfect for weather, where every part of the atmosphere matters as much as every other part. Yet when applied to the universe, most of the grid is wasted because even an entire galaxy is absurdly small compared with the rest of the universe.

Imagine a map of a few cities dotted through a large desert; tourists won't thank a cartographer for devoting equal space to the desert as the cities. The wilderness takes up most of the paper, while the cities are tiny on the page and consequently lack any detail. Likewise, grid-based simulations of the universe spend all their power describing vast wastelands, leaving no room for vital detail in the galaxies.

This problem doesn't affect dark matter which is traced using smarticles—chunks of material that can fly through a simulated space without the need for a grid. Where there is no dark matter, there are no smarticles, and so the computer doesn't waste its time making calculations about empty voids. A similarly efficient strategy is possible for incorporating gas. Instead of using a fixed grid, gas can be bundled into new types of smarticles, a little like the dark-matter ones but feeling pressure forces in addition to gravity. That involves taking the Navier–Stokes equations that underlie weather simulations and re-expressing them to dictate how these new smarticles of gas move.

The first equation concerns conservation and is pretty easy to satisfy: by using a fixed number of gas smarticles, each with a fixed mass, the simulation can guarantee that nothing appears or disappears. Force is a little harder, but still possible to arrange: the computer searches around each smarticle to find the competing pressure and gravitational attraction from its neighbors. The third equation describes energy, demanding that each smarticle keeps track of the heat that it carries and adjusts the pressure it exerts on surrounding smarticles accordingly.

Taken together, the rules give rise to the same kind of complex, swirling motions that I explained earlier for the weather, but the motions are now expressed in terms of moving smarticles rather than a grid. One of the pioneers of the new approach, Joe Monaghan, named it *smoothed particle hydrodynamics*.[23] Monaghan wasn't so interested in entire galaxies, but more in the interiors of individual stars and planets. He spearheaded simulations using the new technique, proving it to be flexible and reliable when applied to the structure of stars and planets, the formation of Earth's moon, and the way that black holes suck material from their surroundings (a topic that I'll revisit in chapter 4).[24] These were detailed studies of relatively small-scale phenomena, but the technique was framed universally: it could focus computer effort where it was most needed for any application.

The broad power of the technique first caught cosmologists' attention in the late 1980s, by which time Monaghan was busy using it to model volcanic eruptions and tsunamis and their possible role in the demise of the Minoan civilization.[25] The approach has found its way into oceanography, biology and medicine, geophysics, Oscar-winning movie special effects, and computer games. Once cosmologists cottoned on, and with the increasing power of computers in the early 1990s, there should have been no barrier to forming simulated galaxies just like the real ones. Instead of a handful of abstract numbers, the new galaxies would be composed of tangibly whirling gas and stars.

These powerful techniques were already being trialed ahead of Hubble Deep Field's 1995 revelations, but the initial results were disastrous, perhaps even contributing to the astronomers' skepticism.[26] Rather than galaxies as we know them with a variety of shapes and sizes, the simulations produced dense, crowded, jostling collections of stars.[27] Compared with those in the real universe, the individual galaxies were too bright and dense. Worse, simulations predicted that a galaxy like our Milky Way would be surrounded by hundreds of mini-galaxies shining brightly, remnants from its long history of merging.[28] These so-called satellite galaxies were far rarer in reality: only a dozen or so had been seen.

By the time I was sitting in undergraduate lecture halls in the early 2000s, there was something of a schism in the community. Cold dark matter had proven its worth, and most cosmologists were convinced it was on the right track. It was certainly a core part of our curriculum. But a significant number of researchers were worried that something was seriously wrong. Dozens of scientific papers issued dire warnings that the foundations of cosmology were inadequate. The improved simulations were supposed to make everything look better; instead, articles of the time spoke about a "cosmic crisis," as the galactic-scale problems seemed to mount without end.[29] One expert on galaxies told *New Scientist* magazine that adherents to the dark universe were "getting pretty fanciful at this point," and that it was time to discard the whole basis of cosmology and start afresh.[30]

The Inescapable Sub-grid

Not everyone was engulfed by this sense of crisis. During a conference in 2005 my future collaborator, Fabio Governato, showed a picture of a galaxy from one of his simulations and declared that all was well.[31] Generating such images involves calculating how light would be generated by the stars, tracing it past gas and dust to check for any shadowing effects, and so finding out how the galaxy would look to a notional distant telescope in the virtual universe. That's worth the extra effort because it allows an immediate visual comparison with reality.

Governato was optimistic that cold dark matter could still make sense of galaxies, and used this visualization of his simulations to show how. At the conference, I remember seeing a disk of stars and gas, swirling around their collective center; it was about the right brightness, and had about the right number of little satellite galaxies orbiting around its central disk. This improvement, Governato said, was the result of steadily improving resolution coupled with a new approach to feedback, which I'll explain in a moment. But to me it still didn't look like much: the disk was fuzzy, vaguely resembling a galaxy but nothing like the spiraling masterpieces that astronomers routinely photographed. The simulated galaxy looked puffy, unlike many real galaxies which are so flat that, when seen from the side, they appear as a razor-thin line of light.

The catered lunch that day happened to be American-style pizza, and I asked Governato—a proud Italian—whether his simulations looked a bit too much like thick-crust monstrosities, rather than the thin-crust beauties they ought to be. With a trace of irritation, he explained that other simulations had only ever managed to produce dough balls. We hit it off immediately.

That year, Governato's simulations were one of a handful to start producing simulated galaxies closer to the real thing. The fuzziness that captured my attention was only to be expected: supercomputers weren't then powerful enough to reproduce the sharpness and detail that can be seen in a photographic image. So the progression from dough balls to a thick-crust pizza counted as a success, and it hadn't been necessary to change or discard dark matter as some were advocating. Instead, it was chiefly about feedback, the effect of depositing energy from the stars' heat and light into the gas.

Including feedback in smoothed particle hydrodynamics simulations at first proved ineffective; even though large amounts of energy from the stars had been included as early as 1992, it had little impact on the resulting galaxies.[32] This was undoubtedly disappointing, and at first it wasn't entirely clear what to make of it. A major motivation for developing the smarticle-based approach had been to get feedback right: when combined with the laws of fluid dynamics, the simulated energy was expected to slow down star formation and resculpt the galaxy. If the energy in fact had little effect, maybe those expectations had been overoptimistic.

In the early 2000s, a growing band of simulation experts—including Governato—started to believe that feedback was important after all, but that the rules governing energy in the simulation needed a rethink.[33] Instead of depositing the energy and using the fluid laws to compute the consequences, they decided to introduce sub-grid rules to maximize the energy's effect. At first, this sounds crazy, or at least unscientific, changing the simulation code because the results had been unpalatable. But, while such thinking formed part of the motivation, there was also a growing realization that no simulation would ever have all the pin-sharp detail needed to capture the interplay between stars and gas correctly.

Stars are around a trillion times smaller than the galaxies they inhabit, and so in reality they inject heat in an extremely concentrated way. In the 1990s simulations, the effects from a star would be spread through its nearest gas smarticle which, even though considerably smaller than a whole galaxy, is still vastly larger than the size of a single star or supernova. The energy was being washed out because the intense, localized effect couldn't be represented by the computer. Governato's team, among a few others, had decided to fix that. Based on rough calculations of the intense heating effect, they worked out how the energy ought to behave within each smarticle, and implemented corresponding sub-grid rules.[34] With these incorporated, feedback started counteracting gravity, making it much harder for new stars to form, just as Larson had said it should.[35]

The need for tweakable sub-grid rules had been obvious when a galaxy consisted of just a few numbers, dating all the way back to Tinsley and Larson's first ventures. But now it was clear that even sophisticated smoothed particle hydrodynamics simulations still needed customized rules because of the inevitable limit to computer resolution. With this successful modification in hand, Governato could show the 2005 conference not just a single static image but a striking movie of how his simulated galaxy was assembled. Like the ultimate time-lapse video, billions of years of history played out over a couple of minutes, giving the whole audience an instinctive feel for the galaxy-building process.

Making one of these movies involves generating thousands of images and playing them back one after another to give the impression of motion. It's pretty much routine for simulation professionals in the 2020s, but back then it was an exciting novelty. The resulting videos can be beautiful, and help the viewer understand what the simulation is saying about history. They typically start with a dark universe in which the lines of a loose, faint cosmic web appear. Then, along this web, little sparks of light ignite as the first stars form. The individual pinpricks grow in size and brightness as the first stars are joined by millions, then billions more. With gravity intensifying, the islands of light are pulled toward their neighbors and merge, building ever-larger galaxies that eventually resemble those of the universe today.

But is this more than a Pixar fairy tale? Just because the simulated galaxies look about right in the present day, and their assembly can be dramatized as an eye-catching movie, doesn't mean that galaxies in the real universe were constructed in this way. Well-established physics dictates the motion of stars and gas; but in the end this provided no escape from troublesome, debatable sub-grid rules. If simulations aren't based solely on reliable, well-established physics, what, in the final analysis, can their movies really teach us?

Learning from Simulations

At its heart, science is not about being correct; it's about having explanations that can be put to the test. My PhD supervisor, Max Pettini, is primarily an observational astronomer but had watched the simulations develop from the sidelines and now encouraged me to compare the history in Governato's videos to reality. The sky is very nearly static over human lifetimes; the cosmos changes only over millions or billions of years, so we can't watch any single real galaxy assemble. But we can at least use the light-travel delay to look back through time, seeing galaxies as they were 1, 5, or 10 billion years ago. That is the point of the Hubble Deep Field, which showed conclusively in 1995 that galaxies have not always looked the same.

Simulations suggest that a regular galaxy like our own Milky Way has assembled from tiny mini-galaxies, the sparks of light dotted along the early cosmic web. The trouble is, all these little fragments are individually too dim to see over the large distances needed to peer back in time—that is the claimed reason the Hubble Deep Field is so empty.

Only the most exceptional galaxies, ones which will grow into giants far larger than ours, can be seen directly, even with space telescopes. Directly verifying the story of diminutive, distant mini-galaxies coalescing is difficult; but without doing so, it is impossible to be sure that simulations are telling the right story.

Pettini is an expert on a neat escape from this problem: he uses telescopes not so much to look for galaxies themselves but for their shadows. The universe is dotted with very bright beacons known as *quasars*. I'll explain more about these in the next chapter, but for now all that matters is their exceptional brightness. Since their beams are visible from across the universe, the light has traveled through all the intervening space and time before reaching us. If it happens to run into a mini-galaxy, even if that mini-galaxy contains very few stars, the gas within creates a shadow.

Helpfully, gas doesn't obscure the light altogether—rather, it blocks very specific hues depending on the galaxy's chemical composition. When astronomers produce a spectrum of the light from quasars, certain tiny bands of color are therefore missing; these missing bands are known as *absorption lines*. The result is a calling card, telling us unambiguously that the light has passed through one or more galaxies on its way through the universe, and giving insight into what's inside those ancient galaxies. We can receive the message even if the galaxies are dim; only the single background beacon need be bright.

The research that I settled on for my PhD investigated whether these shadowy fingerprints of galactic assembly were being correctly predicted by Governato's simulations. For the first few months, I enjoyed working with the simulations, probably because it gave me access to some big computers. I began writing a computer code to predict the absorption lines that galaxies in the simulations would imprint, which could then be compared with the typical lines seen by astronomers.

Meanwhile, I was learning about how the simulations worked in practice and had started to feel skeptical. Dark matter and dark energy were bad enough, although by this time I had reluctantly accepted that the evidence for their existence was pretty strong. Layered on top comes the assumption that gas in the universe can be represented by giant smarticles—at least that is based on the trusty three laws of fluid dynamics, but it still obliterates all the small-scale detail that the computer can't fit. Finally, the sub-grid deals with what the simulation otherwise couldn't, providing an imperfect replacement for everything that's missing, which is then tweaked until present-day virtual galaxies look reasonable. It seemed precarious, and I became hung up on whether simulations mean anything at all.

After thousands of lines of code, but without obtaining any useful results, I was fed up and requested to switch to a different project. I spent the best part of a year on almost unrelated research (which I'll return to in chapter 6), but eventually realized it would be foolish to let the computer code I had been writing go to waste, and so returned to the simulations. My skepticism was such that I expected the results would show a disastrous mismatch between simulation and reality. I had talked myself into the idea that simulations were hokey.

To my astonishment, the simulations actually matched reality. It even explained something which had been bothering astronomers for a while: observations of the distant, ancient shadows showed very little sign of any heavy elements. There was less than a thirtieth of the concentration of carbon, oxygen, iron, and silicon that is present in the Milky Way. This absence of such vital atoms was true in the simulations, too, and we were able to show why: they are manufactured in stars, only a few of which had formed in each of the shadowy mini-galaxies of the early cosmos. Only later do they accrue sufficiently in the assembling galaxies for rocky planets like ours to be viable.

Even now, I can feel the shock of seeing all this work out and make sense. Richard Ellis's 1998 warning that simulation feedback rules could be tweaked to fit observations was correct, and I had taken it to heart. But nobody had been tweaking the rules to reproduce shadows, which, by giving prominence to the most dim and distant fragments of modern galaxies, represented a different side to the galactic creation story. The simulations were therefore free to make a genuine prediction in that arena, a prediction that agreed with and explained reality. It offered some impressive reassurance that simulations, for all their faults, are capable of weaving a story close to truth.

Here is another example from my own experience, when simulations have taken me by surprise. As computer power increases, the resolution of the simulations can also be increased, adding small-scale detail which should make galaxies more lifelike. In 2010 I was visiting Governato in Seattle when he showed me movies from his latest simulations, in which feedback had taken an unexpected turn. The energy that was supposed to keep gas from forming stars had turned violent, and was pushing the gas thousands of light-years out of galaxies. It wasn't because any changes had been made to the sub-grid rules themselves, but because the increasing resolution allowed the explosive force of the stars' energy to have even greater effect.

The real surprise was that, as this gas exploded out of galaxies, it seemed to be dragging dark matter along with it. Ever since the 1990s, astronomers had been worrying that real galaxies, especially the smallest ones, seemed to have slightly less dark matter in their center than simulations predicted; now the new simulations could account for it. A mismatch between simulations and reality, put forward as a reason to distrust dark matter, had seemingly fixed itself. We wanted to understand how.

Nothing had been added to the simulations to change the behavior of dark matter, so it had to be an unanticipated consequence of increasing the resolution with the existing combination of physics and sub-grid rules. Over the course of a week, meeting in a succession of Governato's favored hipster coffee shops, we analyzed the simulations closely, and realized that in small galaxies the gas doesn't just leave: it is pushed out, then it falls back in, over and over again. On each cycle, the gas carries a little more dark matter with it, like a conveyer belt, efficiently excavating the material. We wrote a paper arguing that small galaxies in the real universe must also go through this repetitive cycle of forming stars, blowing out gas, and then sitting idle while the gas falls back,[36] and later observations confirmed our supposition.[37] The processes happened millions or billions of years in the past, but they were predicted by simulations before being observed in reality.

These kinds of results from simulations are what makes them mean something. Nobody should take at face value everything that emerges from a simulation because we have undeniably cheated to make galaxies fit inside computers. But if we can make predictions of something which turns out to agree with the real universe, we can build confidence in the tale the simulation reveals.

This is the story: dark matter drives all creation by assembling small amounts of gas, allowing the first stars to form. Over time, the resulting mini-galaxies collide and merge into progressively larger structures, again guided by the heft of dark matter. Meanwhile, as each generation of stars dies, the growing galaxies accrue an increasing supply of elements like carbon and oxygen. Eventually, there is sufficient material to permit the formation of rocky planets around stars like our own. It seems that without the guiding hand of dark matter in all this—forming the stars, keeping tight gravitational hold of the elements they manufacture—we would not be here. Simulations have taught us that our very existence is contingent on the invisible.

The Mystery of Diversity

Approaching sixty years after Tinsley wrote her thesis, and over eighty since Holmberg darkened his laboratory, galaxy simulations combining aspects of both their pioneering schemes have become routine. As computer power increases and simulation codes improve, we are continually learning more about the islands of light in our universe. Dark matter is their essential ingredient, its weight creating gravitational traps for gas, the fuel for generating stars. When it comes to the gas itself, the three familiar laws of fluid dynamics apply, but must be combined with tricks to concentrate the computer's effort where it is most needed. And, finally, the energy from the stars is carefully traced using sub-grid rules so that the galaxies regulate their own formation. Without that, the computer universes would be dazzlingly bright, in firm disagreement with the sparse, dark cosmos that we inhabit.

Simulations are based on an eclectic prescription: in equal parts dependable physics, computational tricks, and tuning to fit what we already know. Constructing predictions or explanations from the resulting simulations takes care and expertise because of this unusual mix of characteristics. It is all too easy to present simulations as giving us a direct insight into reality, but it should be clear by now that this would be oversimplistic. Disentangling what is a prediction from what is an assumption, what can be trusted from what cannot, takes an expertise of its own and can often be controversial: there are still a very few experts out there who question whether simulations tell us anything at all.

I have some sympathy but think this skeptical take is wide of the mark. Tinsley's trailblazing simulations were able to revolutionize cosmology, long before the mechanisms for growing galaxies and regulating star formation were understood; her masterstroke was to understand that simulations don't have to be literally correct in order to be useful. Modern simulations are still far from a perfect recreation of cosmic history. But they do make predictions about the present and past of our universe, many of which have turned out to be correct. I've given a taste of these kinds of insights from my own experience, which took me full circle from enthusiast to skeptic and back again.

At the very least, simulations prove that the ideas of dark matter and dark energy can be woven into a coherent account of where and why galaxies were born. This story joins the dots between the earliest moments of our universe; the unimaginably vast cosmic web; and the galaxies, stars, and planets that lie within it. Given what powerful telescopes have shown us, it is likely to be correct at least in outline, and this accomplishment shouldn't be taken lightly—it really is rewriting the Book of Genesis, more accurately than Sandage ever dreamed.

But equally, nothing in science is ever completely settled, and many of the ideas that are currently accepted are likely to be revised over time. Finding what doesn't work in simulations is more important than praising what does work: the little cracks in the edifice of modern cosmology provide hope for those inventive theoretical physicists who wish to reinvent dark matter. So far, most cracks have been filled in by improving the sub-grid, but there is no reason why that will continue forever: one day, we may find something which can only be explained by revising the ingredients of our universe.

There is plenty that astronomers and cosmological simulators can't as yet confidently explain. At the top of the list is the sheer variety of galaxies. Some are large and others are small: that's not so surprising, since they scale up and down according to the size of the dark-matter halo that hosts them. Stranger is that some continue to form new stars (our own Milky Way is an example) while others seemingly do not. How did differences like this come about?

The 1995 Hubble Deep Field offered a glimpse of how galaxies change over time, but in modern terms it was a very small enterprise, containing a few thousand of the most exceptionally bright galaxies. Automated telescopes like the Sloan Digital Sky Survey, starting in 2000, have amassed information on millions of galaxies. The galaxies turn out to be diverse in every way imaginable: size, color, shape, mass, chemical composition, age, brightness, rotation speed; astronomers have barely started characterizing the variations, each of which leaves telltale impressions on the light that we can compare with images and movies from simulations. Over the 2020s, the James Webb Space Telescope will show us pictures of the universe even farther back in time than Hubble reached, while the Vera Rubin Observatory (named in honor of the dark-matter pioneer) will gather information about our closest 20 *billion* galaxies.

In spite of how much we already know, these projects could deliver fresh surprises. It's certainly wise to be cautious; as the Hubble Deep Field taught us, it is very hard to be completely sure of what one will find at new frontiers, and it will take time and patience to understand whether and how new data fits with the simulated story. At a minimum, the story of galaxy formation by 2030 will be much richer and more nuanced than the outline we have today.

No two of the 20 billion galaxies that we discover will be exactly alike. The galaxies are built by the same laws of physics, as far as we know, so they must differ solely due to their initial conditions—in other words, they each started out slightly different in the early universe. I'll say more about these differences in chapter 6, but they are tremendously small and subtle. Can we account for how tiny early differences are magnified into the glorious, endless diversity that we see today?

There is a huge array of physical effects that simulations don't yet include which may contribute to ensuring that no one galaxy is quite like any other. Attend a conference on galaxy formation simulations and you will find physicists worrying about the abstruse sub-grid details of magnetic fields, cosmic rays, stellar winds, and space dust.

But, above all else, there is one capricious ingredient that I haven't yet mentioned, which has the potential to destroy some galaxies while allowing others to flourish. I am talking about the universe's greatest known sources of energy, superstars of theoretical physics, delighting elementary-school kids and mathematics professors alike. And, as I am about to show, galaxies don't make sense without them: black holes.

4

BLACK HOLES

In principle, the idea of a black hole is straightforward. It is sup-posed to be a region of space so densely crammed with material that gravity goes into overdrive, pulling and crushing with over-whelming strength. Nothing inside can escape and return to the uni-verse, not even light—and that's why the hole is deemed black.

I loved learning about black holes as an undergraduate; the topic is the ultimate expression of how physics and mathematics can be combined to learn about the universe. In the sanitized history pre-sented to students, Albert Einstein came up with a new theory of gravity in 1915, a physicist named Karl Scharwzschild realized the theory implied the existence of black holes in 1916, and then astron-omers started looking. But in reality, appreciation for black holes took decades to develop, in part because it had to wait for sufficiently sophisticated simulations. It is only now that we are coming to un-derstand the profound effects of black holes on our universe.

Black holes can seem like convenient McGuffins in science fiction, but they genuinely exist. It isn't obvious at first how to demonstrate this; you can't see something which, by definition, doesn't allow light to escape. But it is possible with powerful telescopes to monitor gas and stars in the vicinity of a putative black hole, revealing the immense gravitational forces. Even more tellingly, when black holes collide, they generate distortions in space that spread outward, like ripples from a pebble falling into a still pond. These gravitational waves travel far across the cosmos, and have now been detected passing through Earth.

Consequently, in the last decade, black holes have been demonstrated to exist beyond any last reasonable doubts. The Nobel Prizes in Physics in 2017 and 2020 were conferred on a total of six pioneers who have contributed to the stack of evidence. Two of the six, Andrea Ghez and Reinhard Genzel, found a spectacular black hole, millions of times the mass of our sun, lurking menacingly in the middle of our very own galaxy. Astronomers call it *supermassive*.

Such outsize black holes have gradually become central to my own work because most galaxies seem to cradle one at their center. Where, exactly, they came from is one unanswered question that simulations may help to shed light upon, but in the meantime we know that they can lie in wait for billions of years before suddenly killing the galaxy that nurtured them. It's not so much the direct pull of the hole that is a problem; it is only up close that the hole's gravitational effect is large. Instead, the broader danger of black holes lies in their ability to launch intense beams of radiation, tearing gas from the heart of their parent galaxy, depriving it of crucial fuel for any future generations of stars. (It still takes billions of years for the existing stars to fade, so it is a slow death.)

The way that galaxies and black holes interact with each other is only partially understood. It is a struggle for computers to capture black holes correctly, especially because the radius of a supermassive black hole is around 50 billion times smaller than its parent galaxy. Given the enormous contrast, the only way to include a black hole in a cosmological-scale simulation is as a set of sub-grid rules, just as with stars.

Alternatively, one can temporarily forget the rest of the galaxy and focus all the computer's efforts on one or two individual black holes. In this case, a specialized grid of the right scale can be drawn for the task at hand. Even so, this approach requires cunning tricks to turn Einstein's famously mind-bending theory of gravity, general relativity, into something a computer can handle.

The theory is tried and tested, but it has strange consequences: that time isn't the same for everyone, that matter can pile up in an infinitely dense point, and that black holes have eccentric cousins known as *wormholes*, which theoretically act as shortcuts between different parts of the universe. To take such absurdities seriously required a major leap of the imagination, one that began in the midst of the First World War.

Collapsing Stars

Lewis Fry Richardson, the weather-forecast pioneer, was not the only physicist on the 1915 front line with a grand vision. On the other side of the trenches, Karl Schwarzschild could hardly have been a more different character. Unlike the retiring Richardson, he was outgoing and vivacious, having hosted raucous parties at the observatory in Göttingen where he was director.[1] Far from being a pacifist, he volunteered for military service—by no means necessary, given that he was forty and working in government inner circles. He then served in various roles, including on the front line, calculating missile trajectories.

Schwarzschild was fascinated by stars and by general relativity, which Einstein had perfected in late 1915. By early 1916, Schwarzschild had written two papers using Einstein's theory to describe the gravity around stars, unearthing a strange consequence: an upper limit to how dense a star could become.[2] Schwarzschild reported that if the sun were to shrink less than three kilometers in radius (four-millionths of its current size), no force would be strong enough to hold it up, and therefore such small stars were impossible.[3] This was quick work, especially given the circumstances under which he was operating. It wasn't just the war; he had also developed pemphigus, an autoimmune condition that causes painful skin lesions. A week after his work was published, he died from complications of the disease.[4]

The calculations impressed Einstein, but even he couldn't understand what to make of this lower limit to the size of stars. It seemed strange but probably unimportant; few physicists contemplated the possibility that any celestial body could shrink so dramatically. It would take decades before it became clear that stars really could contract to be smaller than this *Schwarzchild radius*—and that all remaining material would be crushed into their center and nothing, not even light, would escape. In other words, these stars would become black holes.

It seems surprising that such an important conclusion could be missed for so long, but it comes down to the difficulty of understanding equations. On the page, Einstein's general-relativity equations consist of a handful of symbols which look elegant—beautiful, even. But the appearance belies extreme complexity, with each symbol standing for multiple layers of mathematical manipulations that fill entire textbooks. Schwarzschild had solved those equations for the particular case of a stable, spherical star—but that is mainly an exercise in mathematics, and turning it into physical understanding requires extra work. In Douglas Adams's *Hitchhiker's Guide to the Galaxy*, a computer delivers the solution to "the Great Question Of Life, the Universe and Everything." It is supposedly forty-two, an answer that the machine insists is correct even if it satisfies nobody. Solving Einstein's equations can feel a little like this; even if one obtains mathematically watertight results, their meaning can remain obscure because the variables involved are themselves complicated. The mathematics is difficult, but interpreting it is far harder.

The first physicists to suggest seriously that stars really would collapse to Schwarzschild's critical radius, forming real black holes in space, were J. Robert Oppenheimer and his student Hartland Snyder in 1939. At the time, Oppenheimer had ten or so students, all working on different topics, and one of his great talents was to spot unusual and interesting lines of inquiry.[5] He asked Snyder to look into the eventual fate of a star that ran out of energy: seemingly abstruse, but Oppenheimer guessed the answer would have profound implications for theoretical physics.

Normal stars exist in a fine balance between gravity pulling inward and pressure pushing outward, but the required pressures can only be generated by high temperatures, and once nuclear fuel is exhausted the star quickly cools and loses its balance. Oppenheimer and Snyder showed that, in the complete absence of pressure, the star would collapse inward, becoming smaller than Schwarzschild's critical radius. "The star thus tends to close itself off from any communication with a distant observer; only its gravitational field persists," they wrote, describing for the first time the defining characteristics of a black hole.[6]

But this wasn't the end of the matter, because the idea that a star loses all its pressure is an oversimplification. Another of Oppenheimer's students showed that, under the right circumstances, dead stars might instead explode, leaving behind dense but still visible *neutron stars* supported by the pressure of nuclear forces.[7] The true fate of stars hinged on a mire of detail about how different parts of a dying star push on each other; it is the kind of situation best analyzed by a simulation, but this was 1939 and computers weren't yet available.

The outbreak of the Second World War brought work on understanding black holes to an abrupt halt. Most of the experts, including Oppenheimer, became entangled with the Manhattan Project, sometimes in spite of their reservations around the development of nuclear weapons. (Oppenheimer, despite being a key figure, seemed more ambivalent about weapons development than many of his colleagues and was regarded with great suspicion by the FBI.)[8] The existence of the nuclear-bomb-calculating, weather-predicting ENIAC by the end of the war might have enabled the fate of massive stars to be determined, but physicists were now tied up with the start of the Cold War and the rush to develop a hydrogen bomb. As a result, two decades passed before the simulations exploring the fate of massive stars would be attempted—and only then because they, too, became woven into political tensions. In a strange twist of fate, the prospect of nuclear war would necessitate understanding the death of stars.

Black-Hole Simulations

This connection between war and space came to the fore in 1955, when the United States, UK, and Russia began testing thermonuclear devices above the atmosphere, prompting concern over the effects on human health. Stirling Colgate, a nuclear-weapons expert at Livermore National Laboratory, was asked by the US State Department to act as a consultant during negotiations of a test-ban treaty.[9]

Colgate might have had a very different career: his father and uncles had incorporated the famous toothpaste company, which was growing rapidly.[10] But while he was studying at the tiny Los Alamos Ranch School, he became interested in physics. By coincidence, the whole school was bought in 1942 by the US Army because it wanted to develop the land for use as a secret nuclear-weapons laboratory. Colgate knew something was up: he spotted famous physicists who were covertly wandering around the campus, going by pseudonyms but clearly recognizable from pictures in his school textbook.[11] A decade later, he would himself be a key physicist in the project.

In his capacity as a treaty adviser, Colgate had realized that any putative thermonuclear test ban would need to be monitored and enforced. But the explosions of dying stars far beyond our own solar system might generate a flash of radiation very similar to a bomb in the upper atmosphere. These cosmic flashes would be intrinsically far brighter than the bombs, but would be dimmed by the immense distance and might therefore be misidentified as weapons in space, creating false alarms. When he shared this concern with the negotiators "there was great consternation among the Soviet delegation; I mean real consternation. This idea caught them totally by surprise."[12] The deadly consequences of an escalating misunderstanding were unthinkable, so it became crucial to understand how the deaths of stars in deep space compared to explosions just above Earth.

Colgate gathered a team to repurpose existing weapons simulations, since much of the physics was identical. Whether a bomb explosion or a collapsing star, their simulations split the problem into a series of imaginary concentric, nested spheres, tracing how each moves inward or outward while pushing on the others. This strategy of turning a messy three-dimensional problem into a series of perfect spheres risks losing essential details, but even with the powerful computers available to the military it was the only option at the time.

Once the simulations had been adapted, they showed that when a star runs out of fuel its center starts to collapse. So far, this was no surprise, but what happened next depended crucially on the mass of the stars being simulated. If they were small enough, as the nuclei of atoms started touching each other they would bounce back, like trying to squeeze too many marbles into a tiny box and finding they burst out. A massive shock wave would result, thrusting the outer layers of the star outward at nearly the speed of light. This is a *supernova*.

Colgate predicted that these explosions in space would be picked up by military satellites and was proved right, though luckily the precise signature turned out significantly different from that of a weapon.[13] The layers thrown into space glow extraordinarily brightly at first, then fade over time. The crab nebula is a beautiful example of a nearby supernova remnant which, 1,000 years after an initial explosion recorded as visible in the daytime by Chinese and Japanese astronomers,[14] continues to glow, cool, and expand into the surrounding space. The remaining star, with a much lower mass, is tiny but stable: a neutron star, more than 100 trillion times denser than the sun.

In simulations of more massive stars, however, the core continues to contract; even the nuclear forces would not be enough to reverse course. General relativity deviates increasingly from Newton's older theory of gravity when densities become this high; consequently, two of the team members, Richard White and Michael May, incorporated Einstein's equations into the code.[15] But those changes also couldn't stop the collapse, and so it became clear that sufficiently massive stars would be crushed less than a second after the core started contracting. It now seemed that black holes were a natural and inevitable consequence of Einstein's theory.

The adaptations from weapon to collapsing-star simulation were relatively straightforward and went without a hitch; yet it took until 2005 to complete the next step, investigating a collision between two black holes to study the production of gravitational waves. Relativity is an enormous topic, taking years to learn and decades to master, and when simulating more than a single black hole, some of its strangest features loom much larger. To give you a flavor of why designing such a simulation requires a complete command of relativity, I am going to introduce two of its distinctive features: the malleability of time and the existence of singularities.

First, the passage of time, a crucial ingredient of simulations, depends on how a black hole is viewed. The simulation developed by May and White is conceived from the perspective of an unfortunate astronaut falling in along with the collapsing star itself. Seen by a more safety-conscious astronomer at a distance, the simulated scenario looks dramatically different. As the collapsing star shrinks to a few kilometers across—approaching the critical *Schwarzschild radius*—the flow of time itself is disrupted, the collapse appears to slow until it freezes and the star gradually fades from view. There is no outward sign of the infinite crush within. There is just a frozen, dark sphere.

This might sound like some kind of optical illusion, but according to relativity, time actually flows differently for the astronaut who falls into a black hole compared to the astronomer on the outside. It's an effect physicists have demonstrated, on a much smaller scale, by comparing ultraprecise clocks on the surface of Earth to identical clocks that spent sixty hours flying at high speed in aircraft far above our planet.[16] Such experiments confirm that time ticks differently depending on where you are and how you are moving. Accordingly, results from any simulation must be interpreted carefully to distinguish between different possible meanings of time.

This would be confusing enough, but there is an even more troubling consequence of relativity. Consider the falling material in the May–White simulation; nothing can stop it moving inward, and so it piles up in the absolute center. As the star shrinks to become ever smaller, the density and pressure of the material rockets until it can no longer be meaningfully computed. This is known as a *singularity*.

Singularities are Bad News. If you try to fit all the mass of a star into a single point of space, the relevant equations tell you that the density at that point must be infinite. Infinity is very hard for computers to deal with because it doesn't obey the normal rules of arithmetic. At a singularity, an infinite pressure tries infinitely hard to push material outward. It can't succeed, because there's also an infinite resistance from gravity. Yet two infinite forces can't be assumed to cancel each other; mathematicians know that infinity minus infinity is, sadly, not zero. The result is undefined.

This impalpable behavior of a singularity was best illustrated by the Chinese philosopher Han Fei in the third century BCE. He wrote about an arms dealer who claims to sell impenetrable shields alongside arrows so sharp that they can pierce anything. "How about using your arrows to penetrate your shields?" heckles a bystander, flummoxing the merchant.[17] A singularity represents something like an unstoppable arrow meeting an impenetrable shield, and mathematicians can't explain what happens next any more than a sales representative can. As soon as there is a singularity anywhere in the simulation, the rules of arithmetic become useless.

May and White dealt with this by stopping their simulation the moment a singularity appeared: no problem, since their conclusions are all about the microseconds leading up to that defining moment. But today, our best evidence for the existence of black holes come from the gravitational waves which are produced when two ancient black holes fall toward each other and collide. Simulations of these ripples require the computer to be able to represent black holes that have existed for millions of years prior to their collision.

That requires a more sophisticated plan for avoiding the difficulties posed by singularities, something which would take until the twenty-first century to perfect. The solution has its roots in the final bizarre implication of relativity that I am going to cover: wormholes, portals from one part of the universe to another.

Waves and Wormholes

The way modern simulations avoid singularities is delightfully sci-fi. In 1935, Einstein coauthored a paper with his junior research assistant, Nathan Rosen, suggesting that the black hole might be just one half of the story, and that a pair of black holes could act as entrances and exits for a wormhole through space.[18] Draw a line that extends through the entrance of one black hole, and it can end up coming out the mouth of the other, like a shortcut to a remote part of space or even another universe entirely. The singularity, seemingly, is replaced by a mysterious gateway.

Whether any black hole in the real universe could act as a wormhole is doubtful. Einstein suggested no mechanism that would form the entrance/exit pairs, and a collapsing star certainly wouldn't—it makes just one black hole. But Einstein and Rosen showed mathematically that replacing any black hole with a wormhole entrance makes no difference to anything conceivably measurable from the outside, where all that can be seen is a dark sphere. The sphere is known as an *event horizon* because any events taking place inside can have no bearing on the universe beyond: if even light can't escape, there is no way for news to be beamed to the exterior. Whatever happens within the event horizon—wormhole entrance or singularity—stays within the event horizon.

In the 1950s, John Wheeler, originally a nuclear physicist, was the first to suggest that wormholes should be co-opted into simulations of black-hole collisions. Wheeler's interest started from a curiosity about stars, nature's fusion reactors; but when he heard about Oppenheimer's students' work on collapsing stars, he became obsessed with understanding black holes. Wheeler realized that even during an extreme event like the collision of two black holes, the horizons would never reveal what was inside, so he reckoned that simulations should be instructed to regard them as wormholes instead of singularities. This would remove the problematic infinities, and even if the true interior of a black hole behaves differently, the exterior would look just the same. Today, simulators talk about a *puncture* in the simulation, because the singularity is cut out from appearing.[19] Wheeler suggested to his student Richard Lindquist that he simulate two of these punctured black holes colliding.[20]

Wheeler's purpose was to understand the theoretical implications of relativity for their own sake irrespective of whether such implications could ever be tested.[21] But today, we have a more experimental motivation: when two black holes collide, they generate gravitational waves that can be detected passing through Earth. I previously mentioned that the laws of fluid dynamics can be used to describe waves in water, with the surface oscillating up and down, and disturbances rippling outward. Similarly, the laws of general relativity ascribe to space itself a flexibility that behaves rather like the surface of water. Gravitational waves are the result: space can be momentarily distorted by fast-moving, dense objects before springing back into shape. Simulations can reveal the waves predicted by Einstein's theory when black holes collide, provided that the difficulties associated with singularities have been overcome.

Lindquist searched for someone who could code the wormhole trick into a computer, and found mathematician Susan Hahn. She had arrived with her husband in New York in 1951, fleeing their native Budapest after its siege and occupation by Soviet forces. At first, she worked at a bank, but she had always dreamed of becoming a mathematician, and enrolled for a PhD at New York University, first through evening classes and then full time.[22] Hahn's thesis, completed in 1957, delved into the detailed technical challenges of turning all manner of equations into grid-based simulations of the greatest possible accuracy.[23]

Knowing all the possible pitfalls, she was exactly the right person to turn abstract ideas about punctured space-times into concrete calculations. On top of that, she had started working at IBM and therefore had access to plentiful computational power—the corporation was keen to showcase the ability of its machines by tackling hard physics problems. Wheeler's idea to use wormholes as a means to remove singularities was fleshed out by Lindquist, and the way to fit it all into a computer was provided by Hahn.[24]

Hahn and Lindquist wrote that their work proved a "matter of principle" rather than carrying any scientific importance; the experimental search for gravitational waves from black-hole collisions was a long way off.[25] They simulated a head-on collision which is almost impossible in the universe at large: in reality, holes orbit around each other, edging very gradually closer until finally plunging and colliding at a glancing trajectory. More than that, and despite the neat solution to the singularities, the results became nonsensical as the black holes approached each other. This was a reflection of chaos in the system: just as weather forecasts can't predict more than a week or two ahead, initially small inaccuracies in the description of space escalated into major errors before the holes collided.

For this reason, the simulation indeed fell short of solving any practical questions about gravitational waves, but the principle of using a puncture to replace a singularity had been established. It would take another four decades, and many detailed technical insights, before chaotic behavior could be sufficiently suppressed for the entire time it takes for black holes to spiral inward and collide. At last, like buses that failed to turn up for decades, three independent groups almost simultaneously unveiled computer codes that could achieve the formerly impossible, showing black holes spiral together and coalesce.[26] That was in 2005, ninety years after Einstein first wrote down the equations that, unbeknownst to him, predicted this whole new realm of exotic physics.

It would take a further ten years until the first waves from colliding black holes were detected in the real universe in 2015 by LIGO, the Laser Interferometer Gravitational-Wave Observatory. Building LIGO, first envisioned in the 1960s, was itself a technological feat. In parallel, simulators perfected their techniques and amassed a library of the kind of waves that might be detected. By matching to this library, the consortium of hundreds of LIGO scientists could confidently announce that two individual black holes, with thirty-six and twenty-nine times the mass of our sun, respectively, had merged in the distant universe, hundreds of millions of light-years away. The resulting black hole contained sixty-two times the mass of the sun.

You may have spotted that those numbers don't add up, but it's not a mistake: the missing mass was carried off as energy by the gravitational waves. Einstein's most famous formula, $E = mc^2$ (energy equals mass times the speed of light squared), states that such a conversion is possible, and nowhere is this more evident than around a black hole. The total energy in gravitational waves from the last few seconds of spiraling and collision is equal to that generated by all the billions of stars in the Milky Way shining for 1,000 years.

Energy

The comparison between LIGO's detections and simulated gravitational waves confirmed that black holes exist and behave just as relativity's equations predict. But in the mid-twentieth century, when the reality of black holes was far less certain, astronomers had started to notice something awry in the cosmic-energy budget. The first hint came from radio telescopes, which in the 1950s and 60s revealed intense radio signals arriving from space, like cosmic homing beacons.

Astronomers started to point their more traditional optical telescopes in the direction of these radio sources, finding bright pinpricks of light that at first looked just like stars. One of the first to study these dots in the 1950s was Allan Sandage, the same cosmologist who would later resist the idea that galaxies change over time. He was baffled: the colors were nothing like any star that he had seen before, and there was no obvious reason why stars would ever produce intense radio waves. The objects became known as *quasars*, short for quasi-stars; they are the bright beacons that I mentioned in the previous chapter.

Sandage worked at the Carnegie Observatories in Pasadena and discussed the conundrum with his friends a few blocks away at California Institute of Technology. In the end, the puzzle would be solved by three Caltech rising stars and Sandage was cut out from much of the glory, to his fury.[27] The reality sketched out at Caltech was astounding: these dots weren't stars at all. Instead, they were something vastly brighter, but much farther away; the light originated in the hearts of distant galaxies and had traveled across much of the universe. The only thing that anyone could imagine generating so much energy in such a small dot is a supermassive black hole, millions or even billions of times larger than the sun.

Any black hole in isolation is, by definition, completely dark, but when surrounded by gas, it can start to shine brightly. Gas is susceptible to being captured by the black hole's strong gravitational field, spiraling gradually toward the event horizon, forming an *accretion disk*. Individual clouds do not all move in precisely the same way and consequently collide or rub against each other, turning their motion into heat and, ultimately, into light and other forms of radiation. The process is ten times more efficient than a star's nuclear fusion at generating light from mass, and the larger the black hole, the faster it can suck material to maintain its power; the net release of energy can be enormous. This small region around a black hole therefore generates light; from a distance, it is one of Sandage's pinprick quasars, millions of which have now been found, sprinkled like beacons across space.

When the gas is finally swallowed, its spiraling is carried over into a spin within the black hole itself, storing further energy. Despite being hidden behind an event horizon, the spinning black hole's effect on magnetic fields might dramatically expel material from the accretion disk, thrusting it back into the universe at near the speed of light using the stockpiled power, and generating the intense radio waves that first caught the attention of humanity.[28]

There is a black hole at the center of our Milky Way. Imagine what it would be like if substantial material started falling in, turning our own galaxy into a brightly shining quasar, tinged blue in color due to the extreme heat. After a central shroud of dust that currently obscures the galactic center burns away, an eerie new light would illuminate the night sky, shining 1,000 times brighter than Venus. Such a source would be visible in the sky even during the day, despite being more than a billion times farther away than the sun.

The human race wouldn't be threatened by any of this, but the long-term future of the galaxy might be. Nowhere is this more apparent than in M87, a nearby elliptical galaxy which is disturbed by a jet of material reaching directly outward, thousands of light-years from its center, traveling at almost the speed of light, pummeling anything in its path, like a true-life Death Star. The beam is narrow, so it's not likely that individual stars or planets would be destroyed. On the other hand, this intense energy has to go somewhere—so what happens to it?

Black Holes in Galaxies

Developing simulations that can explain what happens to the energy from black holes is an audacious project. It isn't a case of bringing together detailed simulations of black holes with existing simulations of galaxies, because there is a huge mismatch in scale. The most massive known black holes are estimated to contain many billions of times the mass of the sun—yet their event horizons are still only the size of a single solar system. Galaxies themselves are thousands of times more massive and tens of billions of times larger. The size mismatch is of the same proportions as a global weather forecast tracking an individual speck of dust. This is out of the question; the only way forward is to use sub-grid rules which, by the early 2000s, were starting to help galaxy-formation physicists include the effects of stars in their simulations. Astrophysicist Tiziana Di Matteo became convinced that a similar approach could be applied to black holes, too.

Di Matteo had completed her PhD in Cambridge researching how matter falls into black holes and generates energy. At first, she was primarily interested in using radio, optical, and even X-ray telescopes (which receive the same kind of energetic radiation that's used to look inside your body) to see how the real universe worked. But she was working at Harvard University alongside Lars Hernquist and Volker Springel, both of whom were pioneers in improving the sub-grid.[29] Sensing an opportunity, Di Matteo persuaded them to extend their rules to include black holes.[30]

The mind-bending distortion of space and time is crucial if you care about precisely how the energy and gravitational waves are released, but, the trio hoped, the rest of the galaxy wouldn't care about such details. They envisioned a black hole as simply another type of simulation smarticle, adding to the dark matter, gas, and stars already present. A black-hole smarticle follows one specialized rule: if immersed in gas, it gorges on it. As it does so, it converts a fraction of the mass into energy. As with the rules applying to stars and rain clouds, there are then details to worry over: Just how quickly can a black hole consume gas? What fraction of the energy is released? In what exact form does the energy emerge? There are no definitive answers to these questions, even today.

It was the team's willingness to take a stab in the dark that allowed them to plow ahead with the first simulation of the resulting effects in 2005.[31] While that is the same year that detailed black-hole-gravitational-wave simulations began to work and that cosmological simulations began to form more realistic galaxies, the conjunction is largely coincidental. Di Matteo's simulation has most in common with Holmberg's from six decades previously, in the sense that it involves two ready-made galaxies thrown toward each other, with the intention of seeing what happens during a collision. Instead of Holmberg's thirty-seven bulbs, early twenty-first-century technology allowed each of Di Matteo's to comprise 30,000 simulated particles of dark matter, 30,000 stars, 20,000 smarticles of gas—and one supermassive black-hole smarticle 100,000 times the mass of our sun.

The galaxies would take around a billion years of virtual time to coalesce, and the team took a snapshot of their galaxies every few million years. That allowed them to create a dramatic animation of the results, just like the ones Governato showed of his galaxies forming from the cosmic web. Working late one night, frantically preparing for a conference talk, Di Matteo viewed the results for the first time—and immediately realized her team were onto something exciting.[32]

In the video, two unblemished disk galaxies head toward each other across an empty void of space. As they approach and touch, gas from the two galaxies is squashed and forced into their respective central black holes, liberating vast amounts of energy, which heats the surroundings. The galaxies appear to be on fire, billowing smoke at the camera; in reality, this is superheated gas from the vicinity of each galaxy's supermassive black hole. The smoldering galactic embers start to calm a little; they are pulled together by gravity and coalesce in the middle.

Like the best disaster movie, just as it seems to be over, it gets worse. As the newly amalgamated galaxy begins to settle, the two black holes reach its combined center where some gas remains. They start to consume matter once again, and the fire is explosively renewed; whatever doesn't fall into the holes is driven relentlessly outward. While the existing stars are small enough to escape any damage, the loss of so much gas still devastates the galaxy since there is no fuel left to form future stars and planets. The old stars will eventually fade, and so these two disk galaxies have been killed by black holes, leaving a single, dead remnant.

The idea of destroying entire galaxies in this way would seem fanciful if it weren't for the clear evidence seen in the real universe for the power of black holes.[33] The picture that the new simulation presented helps explain the relationship between black holes and galaxies: it had been observed for some time that larger galaxies host bigger black holes.[34] As a galaxy grows through successive mergers, so, too, must its central black hole, which eventually becomes so powerful in relation to the stars that it can turn on its keeper. In the mid-2000s, simulations had only just discovered how feedback energy dictates the relationship between dark-matter halos and the galaxies they host; now they showed the same synergy between galaxies and the black holes within.

More recent simulations paint a subtler picture in which the gas in a dying galaxy is lost more gradually, rather than in a single calamitous encounter, but black holes are still seen as the engines of destruction.[35] There remain profound puzzles to be addressed, not least where these enormous black holes come from in the first place. Di Matteo initially added them by hand, but if cosmologists really want to understand the story of galaxies, we now need to know what governs the birth of supermassive black holes. The relatively small holes that result from a stellar implosion could not grow quickly enough to match today's supermassive sizes.

At the moment, our galaxy formation simulations are programmed to create supermassive black holes in the centers of young galaxies, without a strict justification for doing so. One possibility for what happens in reality is that the first generation of stars are huge, perhaps 1,000 times more massive than our own sun, and generate correspondingly enormous black holes that consume their surroundings quickly. (The biggest stars today are featherweights by comparison, though still weighing well over a hundred times more than the sun.) Another possibility is that, in the early universe, conditions permitted gas clouds to collapse under their own gravity without ever igniting nuclear fusion. This would bypass the stellar stage, leading naturally to enormous black holes. A third possibility is that black holes somehow start forming long before galaxies themselves. We don't know which theory is right; it is a mystery for future simulations to explore.[36]

The Future

Whatever the exact mechanism, observations show that black holes are present in galaxies, and so we put them in the simulations, too. That has a natural consequence: if two galaxies merge, the resulting single galaxy ends up with not one but two supermassive black holes. And remember that cosmologists have very good reasons to believe that most galaxies, including our own, have been built by successive merging of multiple mini-galaxies over time. So, it shouldn't be a huge surprise that in recent simulations that my collaborators and I completed, the Milky Way doesn't have just a single, central supermassive black hole—it could easily have a dozen, give or take.[37]

Most of the dozen supermassive black holes in our simulated galaxies are not at the center but orbiting far out. There is little gas for them to swallow, so they don't shine or grow much and they are likely to be exceptionally hard to detect. Although it sounds dangerous to have these hidden monsters floating around, the galaxy is an absolutely enormous place and the chances of one coming even within a few light-years of our own solar system are absolutely minute. (I estimate the odds at around one in a billion over the entire lifetime of the sun. The uncertainties are such that the precise number can easily be different, but the chances are definitely very small.)

In fact, we would have found far more of these drifting supermassive black holes in simulations except that they often stray into the middle of their host galaxy, whereupon they merge with the existing central black hole. This is exciting because it gives cosmologists testable predictions for gravitational wave detectors: the regularity with which supermassive black holes collide controls the number of ripples that will pass through Earth.

So far, so good. Humanity has gravitational wave detectors. We have increasingly specific predictions for the waves reaching Earth. Together these should allow us to test ideas about black holes, galaxies, and their tense symbiosis. But, unfortunately, LIGO isn't sensitive to the right scale: detectors must be comparable in size to the black holes they look for. LIGO's detectors are around four kilometers across, making them the same size as a black hole with mass a few times larger than the sun. Supermassive black holes, millions of times more massive, are also millions of times larger in extent. And that means engineers need to scale up LIGO.

Since Earth is only a few thousand kilometers across, there is no room for a multimillion-kilometer detector down here. Flying in space is required to reach such a spectacular scale. In 2037, the European Space Agency (ESA) plans to launch LISA—the Laser Interferometer Space Antenna.[38] It's LIGO enlarged by a factor of a million. No one is building a spaceship so gargantuan; rather, LISA consists of three separate craft, each around three meters across, flying in a 5-million-kilometer equilateral triangle formation. Each shines a laser at the other two, and by monitoring the laser light, the ripples in space between the craft are inferred. Making that bold concept work in reality is at the very limits of humanity's engineering capacity, but a technology test in 2015 convinced ESA that it can be done.

If you can't wait until the 2030s to find out whether simulations are right about how galaxies and their million-times-solar-mass black holes operate, don't worry. Sometime in the 2020s, it's very likely that astronomers will see the first hints of gravitational waves from these giants using a detector that nature already built for us: *pulsars*.

Pulsars are types of neutron stars that spin exceptionally fast, causing them to act a little like a lighthouse, whipping a radio beam around the universe in a fraction of a second. As seen from Earth, they look like a regular pulsating signal—so regular, in fact, that they seem artificial, and on first discovering them in 1967, then PhD student Jocelyn Bell Burnell jokingly labeled them "Little Green Men."[39]

They are not alien, but the astounding regularity of pulsars makes them sensitive to waves rippling across vast distances within our galaxy. As a gravitational wave comes between us and the star, the distance grows and shrinks slightly, and so the timing of the received pulses changes. This effect becomes particularly noticeable when monitoring multiple pulsars, and astronomers have started linking telescopes into *pulsar timing arrays* to facilitate the search. While LISA is needed to confirm whether simulations are exactly right about the relationship between galaxies and their black holes, pulsar timing arrays should give us a first hint much sooner.

The Other Singularity

As cosmologists await these results, there is plenty to ponder about black holes. We now know they are the darkest objects in the universe, yet responsible for generating vast amounts of light and energy, unleashed on a galaxy when it becomes too massive. They are impossible to see, and yet announce their existence with ripples in the structure of space, which can be studied by gravitational wave observatories. And they can be incorporated into simulations by using wormholes to hide away their central singularities which would otherwise violate basic arithmetic, or alternatively by turning them into innocuous smarticles that generate energy through a handful of subgrid rules.

Hiding the singularity at the heart of black holes is a necessity, otherwise simulations would break in the face of the paradoxical infinities. But the wormhole fix is a sophisticated Band-Aid which, while mathematically sound, does not reveal the physical heart of a black hole. To theoretical physicists, the question of what *really* happens deep inside a black hole holds huge sway, even if any implications for understanding the galaxies that surround them would be subtle. The presence of singularities means that relativity is broken, or at least incomplete, and it is only by a stroke of fortune that simulations of black holes have nonetheless proved tractable.

The singularity problem becomes even more conspicuous and pressing when cosmologists consider space as a whole, because the universe is getting larger. The equations of general relativity can extrapolate this expansion backward in time and, in just the same way that a black-hole singularity is the seemingly inevitable end point of material collapsing, a Big Bang singularity is the seemingly inevitable start point of material expanding. This intimate link between the nature of the Big Bang and the centers of black holes was first spotted by Stephen Hawking.[40]

To my mind, Hawking's result is far more troubling than all the exotic physics of black-hole singularities, because the Big Bang cannot be wished away by fancy wormhole mathematics or hidden behind a horizon. Attempts to dismiss the nature of the Big Bang as a philosophical or religious matter aren't of much practical use or comfort. Remember the central role that initial conditions play in all our simulations: you can't predict the weather tomorrow without knowing the weather today. Similarly, you can't make predictions about the universe today without adopting some kind of starting point.

Even if a simulation were to treat the physics of dark matter, stars, gas, and black holes with supreme precision, it could get completely wrong answers (or no answers at all) by making the wrong assumption for what happened in the immediate aftermath of the Big Bang. And so, if astronomers still want to understand why the Milky Way looks like it does, while other galaxies look so different; if we want to understand why some galaxies are large and others are small; if we want to understand why some have been killed by their black holes while others have survived; if we are going to claim that we have simulated the origin of all structures in our cosmos, we have to understand how to treat that initial singularity, replacing it with something more meaningful. That is the mission for the next chapter.

5

QUANTUM MECHANICS
AND COSMIC ORIGINS

It is tempting to introduce quantum mechanics as physics of the microscopic realm. Its bizarre central assertion is that subatomic particles are not solid but fuzzy, able to exist in two or more places at once; plenty of experiments have confirmed this counterintuitive behavior. By contrast, I need only look around me or at the night sky to see that objects in our everyday world and the broader universe occupy single, clearly defined locations at any given moment. It would seem that any fuzziness must be restricted to small scales that we can't see with the naked eye.

Yet if there is one thing I want to relay in this chapter, it is that quantum effects are by no means confined to microscopic, invisible, forgettable little atoms. In fact, quantum phenomena shape and give meaning to the entire cosmos. Current thinking holds that every structure in the universe—the cosmic web, the dark-matter halos, the galaxies, the black holes, the planets, life, you, and me—owes its existence to quantum uncertainty at the beginning of time. Our seemingly solid existence is one facet of a universe that is secretly in-decisive and fuzzy on every scale from microscopic to cosmic. This audacious final piece of the physics jigsaw must somehow be incorporated within simulations.

Quantum mechanics' foundations, absurd though denying a solid reality would seem, are not in any doubt: you need only pull out your phone or tablet to benefit from the resulting technology. These devices are packed full of transistors—digital switches, using one electrical signal to determine whether or not another electrical signal can flow. Such an automated switch allows rudimentary logical reasoning, and chaining millions or billions together has powered the information technology revolution. Transistors are fabricated from semiconductors, materials that harness the quantum fuzziness of electrons to behave in part like an electrical conductor and in part like an insulator. There is no analog in everyday experience that can be used to explain satisfactorily how these semiconductors work; only a knowledge of quantum mechanics will suffice.

But developing such an understanding isn't necessary for comprehending what transistors do, only how they work. Computers do not exhibit any obviously quantum properties: as I am typing into my computer, its memory is filled with particular letters and words. Likewise, simulations predict definite outcomes, like a certain wind speed, amount of rain, or number of stars; to incorporate uncertainty is frustratingly hard, requiring scientists to perform multiple simulations, bracketing the possible scenarios. So the operation of transistors is concrete and predictable; uncertainty is alien to their purpose, even if it is actually key to their internal functioning.

That separation of internal, fuzzy microscale physics from external, predictable large-scale behavior is the argument for describing quantum mechanics as small-scale laws that we don't need to worry about in our macroscopic lives. It's the temptation that I started with: one might more easily accept strange quantum behavior of tiny particles if our everyday world seems safely removed from the weirdness. To show why this clean separation is fallacious, we first need to go further into the microscopic realm, to understand what it really means for a particle to exist in more than one place at the same time.

The structure of atoms, the nature of chemical elements and their grouping into a periodic table, and the way in which molecules bind together all crucially rely on quantum mechanics. Chemists have their own way of simulating the relevant physics, which bears close study because it overcomes immense computational challenges. With those challenges clarified, we will journey back to the first tiny fraction of the universe's existence to see how reality emerged from a quantum froth, to understand what that means for cosmic simulations, and to comprehend why our definite lives are a facet of random chance.

Uncertainty

The most ubiquitous evidence for quantum mechanics is the existence of stable atoms, the building blocks of our familiar lives. Experiments in the opening years of the twentieth century established that atoms consist of electrons orbiting a nucleus, but there was a fundamental problem: according to the existing laws of electromagnetism, the orbits wouldn't be stable. Instead, electrons would spiral into the nucleus over the course of about one hundred billionth of a second, contradicting pretty much everything we know, including the existence of stable, long-lived materials in the world around us.

In 1924, Louis de Broglie, a French aristocrat who had spent the

First World War developing radio communication systems (and installing them on the Eiffel Tower),[1] suggested as a solution that the electron might be smeared, forming a ripple around the nucleus rather than an orbiting particle in the traditional sense. He showed that such a configuration would be stable, bypassing the spiraling problem. The chair of the Nobel Committee, on awarding de Broglie the 1929 Physics prize, sounded almost critical when highlighting that this imaginative leap had been achieved "without the support of any known fact." Luckily the experimental facts came later and showed that de Broglie's idea was spot-on.

Photographs provide a helpful analogy. They cannot be captured at a single instant but instead are built by a camera over a period (albeit a short one, normally a fraction of a second). The result is that fast-moving objects don't appear sharp; instead, the picture shows a smeared version of the object in the different locations it occupied during that time.

With a fancy camera, it may be possible to reduce the blurring by choosing a faster shutter speed, but that's not always desirable. Sometimes blur helps give a sense of motion, so that you can visually gauge from a single snapshot the way objects are shifting over time. Think about nighttime scenes where cars turn into a streak of taillights or sparklers spell out words; they are far more evocative than a perfectly sharp picture. A slow shutter gives that sense of motion while making it impossible to determine a single precise location. A fast shutter captures the position but eliminates any sense of motion. There is a tradeoff.

Quantum mechanics holds that a similar trade-off takes place in reality; as though reality itself is a photograph, information about speed and location is inextricably intertwined. The blurriness in fact takes the form of a wave, rippling as well as smearing—for this reason it is known as a *wave function*. But that is just a detail: if you can accept that nature is blurry, you are on the way to understanding why quantum mechanics matters so much. One consequence is the *Heisenberg uncertainty principle*, which states that position and motion cannot be determined simultaneously.* An experimental physicist can precisely measure one or the other, but not both, very much evoking the photographer's choice between long and short exposures.

Our everyday life seems to flatly contradict Heisenberg; it seems self-evident that, as I drive along the road, my car has a well-defined position and speed at any given time, whatever a photograph might later show. But, rather as with relativity, the hardest step in understanding quantum mechanics is to allow preconceptions about what is likely or reasonable to fall away. Just as relativity only becomes apparent on large scales or at high speeds, the effects of quantum mechanics only become apparent over microscopic distances. When electrons orbit a nucleus, for example, de Broglie showed that the blurry ripples occupy the size of a single atom, around a billionth of a meter.

Simulations of Materials

The familiar world around us is made from molecules: chains of atoms bound together by fuzzy electrons smeared around the entire structure, acting like a unifying envelope. Chemists need to simulate these molecules for all kinds of reasons. They might wish to design a new battery, or a drug, or study the way that viruses attack human

---

* Werner Heisenberg worked for the Nazis during the Second World War, applying his quantum expertise to nuclear energy. Luckily for the world he was more interested in power generation, and decidedly lukewarm about the prospects for developing a nuclear bomb.

cells, or the way that asphalt roads behave when damaged, or search for potential applications of brand-new materials like graphene.[2] Whatever the particular purpose, such simulations tend to keep track of thousands of atoms, or millions in biological systems like viruses.

These *molecular dynamics* simulations are at heart similar to Holmberg's light-bulb galaxies—but tracking atoms instead of stars, aiming to shed light on molecules rather than galaxies. From some initial configuration, specifying the location and movement of the atoms, the simulation allows each to drift for a small time. While drifting, atoms move in a straight line at a fixed speed. The steps through time, that for galaxies are measured in millions of years, are now nanoseconds: molecules can change extraordinarily quickly because they are so incredibly small. After each period of drift, atoms are kicked onto a new trajectory by the forces between electrons and nuclei. The idea of a repetitive drift-kick cycle is therefore identical between molecular dynamics and dark-matter simulations.

The difference is in the particular type of kick. Holmberg worked out gravitational forces by measuring the brightness of light; in modern astronomical simulations, the forces result from digitally summing the tug of all the stars, dark matter, and gas in the virtual universe. But in the case of molecules, quantum mechanics makes forces much harder to calculate: they depend on where electrons are, but, as I have said, each one is in multiple places at once. They are smeared throughout the molecule, and so their forces can only be calculated by a simulation which takes this quantum mechanical effect into account.*

---

* In principle, this smearing applies also to the central nucleus of each atom, but de Broglie correctly suggested the effect would be small, because the nuclei are far more massive.

Some of the first quantum chemistry simulations were inspired by the particular difficulty of understanding how biological molecules behave. The mechanism of vision, for instance, is dependent on a molecule known as *retinal*; however, until the 1970s, it was unclear exactly how retinal turned light into neural signals. The Harvard biologist Ruth Hubbard had shown that there was only one plausible mechanism via which vision could be achieved: quantum effects must turn the light into motion in the retinal, and the motion could then be turned into a nerve signal to the brain. But as to how, precisely, this happened, it was not possible to say—it could not be computed by hand.

Lacking computers that were powerful enough to find the answer, and exasperated with the male-dominated world of science, Hubbard turned her attention to feminism, becoming famous for her excoriating critique of the racial and gender biases within evolutionary biology.[3] (She commented to the *Boston Globe* that she had "no idea what my colleagues think of me. I think at best they're puzzled. At worst, they think I've gone off my rocker.")[4] As a result, Hubbard's exploration of retinal was taken up by her students, among them Martin Karplus.

Karplus would eventually win the 2013 Nobel Prize in Chemistry for simulations, including those of retinal, but he took a circuitous route to get there. Having worked with Hubbard while he was an undergraduate, in 1950 he was unable to continue investigating retinal for his PhD thesis despite wishing to do so. His graduate research supervisor, Max Delbrück, wasn't keen; in fact, Karplus says that when he gave a seminar outlining his ideas, Delbrück repeatedly interrupted to say that it made no sense.[5] The theoretical physicist Richard Feynman was also in the audience and lost his patience with

Delbrück's interjections, whispering loudly, "I can understand, Max; it makes perfect sense to me." Delbrück walked out in a huff, the event collapsed in acrimony, and Karplus found another research supervisor who set him to work on a different topic.

Karplus became an expert on using computers to simulate chemical reactions, and especially on simplifying the effects of quantum mechanics to fit inside the rudimentary machines of the time. But by the 1970s, he was bored: "I had grasped what was going on in elementary chemical reactions and the excitement of learning something new was no longer there."[6] So he returned to the problem that Hubbard had set and that Delbrück and Feynman had sparred over.

Karplus and his collaborators gradually worked out how to represent smeared quantum electrons and their effect on the retinal molecule. Without quantum mechanics, individual electrons would be just like Holmberg's light bulbs, described by their position and motion. This takes six numbers (three to describe the location in space, and three to describe speed and direction of motion). But with quantum mechanics, there are endless possibilities for where an electron *might be*, rather than a single truth about where an electron *is*.

Imagine dividing the space around the molecule into a weather-simulation-style grid.[7] The simulation can now describe the chance that the electron is in each box in the grid, but this requires a number for each. On top of that, because it is not just a fuzz but also a wave, there is an additional number, known technically as the *phase*. If the grid contains a hundred boxes, for example, it will be necessary to store and manipulate 200 numbers to describe one electron.

So far, this doesn't sound so different from the challenges of simulating weather. Things get really difficult only when the simulation needs to track more than one electron. The chance of an electron being in any given box depends on the chance of another electron being in neighboring boxes, an example of a broader quantum property known as *entanglement*. To take this into account, a simulation would need to store two numbers for each pair of boxes, and there are 10,000 ways of pairing up a hundred boxes; it makes 20,000 numbers in all. That's just two electrons in a hundred notional boxes; the whole problem quickly spirals out of control for larger, more realistic situations.

The trick of simulating quantum molecules is all about cutting out this complexity, and luckily that's often possible. For many problems, the overly complicated physics of entangled electrons can be accurately approximated by much simpler techniques, and many electrons can even be ignored entirely. There are almost 160 electrons in a retinal molecule, but most of them lead rather boring lives, orbiting close to the nucleus of the particular atom to which they are attached. A simulation that traced each of these electrons' quantum behavior is unnecessary; as de Broglie pointed out, the nuclei are very massive and the electrons smear around them in a well-understood manner. The simulation needs only track electrons which are far from any single atomic nucleus; these spread into a cloud which can encompass large parts of the molecule, binding it together and determining its shape.[8]

Karplus's team showed that the retinal molecule can be understood by treating only such special electrons as quantum—still a huge challenge, but achievable once combined with computational tricks to speed the calculation. When light strikes the molecule, the electrons receive an energy boost and respond by changing the shape of their quantum fuzz; this in turn pulls the entire molecule into a different shape. The resulting motion creates a chain reaction culminating in electrical signals that your brain interprets. These first quantum simulations completed the picture that Hubbard had sketched decades before.

Inventing tricks to simplify simulations is the name of the game, and plenty of Nobel Prizes have been awarded for it. But they can't overcome the challenge entirely. Far more satisfying would be to find a way of simulating quantum mechanics on a machine which could directly tackle the spiraling requirements, storing a true-to-life representation of entangled fuzz without breaking a sweat. And that might one day be possible—by making computers themselves quantum.

The Promise of Quantum Computers

Simulating quantum physics is challenging because even individual particles are not simple: they spread into a haze known as the *wave function*, representing an irreducible level of uncertainty in reality itself. Tracking the haziness costs time and storage space that quickly mounts, meaning that even large increases in computer power only allow for a small increase in the size of a molecule being simulated. Quantum computers are worth studying both for their potential to smash through this barrier, and because they shed further light on the nature of the strange theory.

Traditional computers are profoundly ill-adapted to uncertainty. Just as the letters on the page in front of you appear in a specific, inarguable order, everything inside the computer's memory contains something particular, like an *A* or a *B*. But quantum physics forces us to represent situations in which uncertainty takes center stage, where all we can say is "possibly *A*, or possibly *B*."

Definite letters and words can still be used to describe uncertain situations. I'm doing my best in this chapter, using concrete language which (I hope) carries unambiguous meaning—despite describing a physical world full of ambiguity. Similarly, a computer can store things in its not-at-all-fuzzy memory that nonetheless represent fuzziness and uncertainty, but it is cutting against the grain. It would be far simpler, if we're not sure whether something's A or B, to have the computer store a symbol for the combination of the two possibilities ($A\!B$).

The idea behind quantum computing is to build a machine from components that, unlike traditional computers, can store and manipulate such fuzzy symbols. Since reality behaves in a fundamentally quantum way, this should be possible; current computers actually do not take full advantage of the physical capabilities of the electrons they use. The most important thing is that the new, purpose-built simulation machines should include effects of entanglement, the connection between different particles that becomes prohibitively hard to track on a classical computer. With the fuzziness and interconnectedness already built into the hardware, once the simulation starts the familiar kick-drift cycle to push forward in time, the final results will incorporate quantum effects automatically.

Theoretical foundations for this idea were laid in the late 1970s, and came to broad attention during a 1981 conference at Massachusetts Institute of Technology, when Richard Feynman delivered the keynote address, conjecturing that such machines would be the perfect tool to simulate quantum systems.[9] The electrons around a molecule give one example of such a system, but the idea was to build a machine capable of simulating absolutely anything where quantum effects are important.[10]

Feynman is a cult figure among physicists, revered for his ground-breaking insights and extraordinary ability to bring science alive through lucid writing and lecturing. Hero-worship is a dangerous thing: his own writing, stuffed with casual yet carefully crafted anecdotes about his intellectual brilliance—and about his shameless attempts at womanizing—makes clear he was a narcissist and misogynist, even by the standards of the day.[11] Still, his ideas on physics are undeniably exciting, and inescapable when one works in the quantum realm. His conference conjecture about the importance of quantum computing machines helped inspire a generation of physicists to take the possibility seriously, despite the huge technical challenges of building one for real.

One such person was Seth Lloyd, who would extend Feynman's conjecture by outlining the design of a quantum computer.[12] "Design" here is a loose word: he explained the kind of machine that would need to be built, without giving detailed blueprints for how to engineer one. It is still unclear today just how feasible large-scale quantum computers will be. But Lloyd demonstrated that in principle Feynman was correct: a single well-designed machine could be repurposed to simulate any conceivable physical scenario where quantum mechanics enters. The precise hardware is almost irrelevant—it can be based on atoms, light, superconducting metals, or anything else that exhibits quantum behavior. This transcendence of hardware echoes how Babbage, Lovelace, and Turing's notion of a traditional computer did not rely on electrical circuits, steam-driven cogs, or any other specific technology; their vision was for a general-purpose machine that can perform any calculation whatsoever, by repeatedly applying a small number of logical manipulations.

There is nothing a quantum computer could achieve that, in principle, could not be calculated by one of these idealized classical machines. It is just that, in reality, all machines have limited amounts of memory and operate for a finite length of time, meaning they can complete only limited numbers of manipulations. The complexity of quantum physics means that practical limits can be reached even for simulating simple molecules. That is why chemists and biologists are excited about quantum computers which, through using quantum effects directly, have the potential to break past these limitations.

In spite of all this promise, construction has proved fiendishly difficult and drawn out. Only recently has it become possible to perform some basic chemical simulations on a real quantum computer (built by Google).[13] It is a feat of engineering: a beautiful machine that looks like the fevered imaginations of a sci-fi set designer, involving a meter-tall series of gleaming metal platforms suspended vertically from above, with bundles of neatly coiled cables and pipes between each one. The vast majority of the apparatus is an elaborate freezer, allowing the core of the apparatus to reach fractions of a degree above the coldest possible temperature, absolute zero. Quantum computers are so delicate that their operations can easily be disrupted by heat, so it is here that the calculations are undertaken, inside a device no larger than a regular computer chip.

The machine is *noisy*, which is to say that it makes errors in its calculations because the technology is so hard to perfect. Such noisy computers are still useful for certain limited simulations, but they do not fulfill Feynman's vision. Noise-free quantum computers remain at the design stage, and experts differ on the timescale for achieving them in practice. Decades might be optimistic.[14]

Still, sooner or later, it will likely become possible to simulate far larger molecules than can be attempted with even the largest classical machines. And, eventually, engineers might even construct quantum computers that do away with the need for all the elaborate cooling, opening the potential for everyone to have one in their pocket.[15] It's unclear whether there is a need for these machines beyond a few specialist applications, but it wouldn't be wise to bet against it: think about our cell phones, which have their roots in room-sized military apparatus of the 1940s. The impossible has a habit of becoming possible, the possible of becoming desirable, and the desirable of becoming ubiquitous.

In my own profession, nobody expects that quantum computers are about to revolutionize the business of simulating the universe. But, equally, it would be wrong to imagine that the universe is immune to quantum fuzz. In fact, our best theories of what happened in the cosmic opening fraction of a second suggest that the entire universe is just as fuzzy and uncertain as Karplus's electrons. It seems hard to believe that the solid reliability of planets, stars, and galaxies could be an illusion, but that is exactly what one bold graduate student—Hugh Everett III—claimed in 1957. Increasingly, many cosmologists believe him.

Quantum Realities

For most of his career, Everett was engrossed in performing simulations of nuclear war—not the individual bombs but the broader strategy of where and when to strike. As part of an elite team of mathematicians and physicists employed by shady arms of the US government, Everett concocted digital realities with death and destruction on unimaginable scales. On the basis of these simulations, many in Everett's circle argued for real-world preemptive strikes on the USSR, not because the outcome would be good for the West but because it would be worse for the Soviets. Luckily, they never convinced politicians, but Everett clearly had an ability to detach from reality in an almost inhuman way. At the time of his death in 1982, he left express directions that his wife throw his ashes into the trash can.[16]

Back in 1957, Everett was a young PhD student working alongside the wormhole-aficionado John Wheeler, and busy trying to understand the implications of quantum mechanics for the universe as a whole. If molecules, atoms, and subatomic particles are governed by quantum physics, the universe they inhabit and sculpt must also be affected by the same laws. One might hope that quantum effects have limited impact on large-scale phenomena, like in the case of a transistor. But I can now explain why this comforting line of reasoning, in which bizarre and unnerving effects are confined to microscopic scales, doesn't quite bear scrutiny.

Imagine a hypothetical far-future weather simulation that tracks the countless molecules within the atmosphere but ignores quantum mechanics, and suppose you edit the simulation to shift a single molecule into an alternative position. The change will have negligible effect at first, but recall Edward Lorenz's butterfly wings from chapter 1: initially tiny differences can be amplified until they determine the entire weather pattern. A molecule is far smaller than a butterfly's wing, but that just means it takes longer for its effect to be amplified— it still has the power to change the far future. So, different versions of this simulation would make differing predictions for the weather after a month or two; in some, perhaps a hurricane hits New York, while in others, it misses, or doesn't form at all.

That's just another way of thinking about the difficulty of making forecasts, but quantum mechanics adds a huge twist. According to de Broglie and Heisenberg, individual molecules do not actually have perfectly well-defined positions. It is not just that nobody knows the positions: rather, their positions all have an intrinsic degree of smudginess. And if a molecule starts out smeared across different locations, all possible different effects on the weather must play out in tandem, at the same time. A hurricane can simultaneously hit and not hit New York.

It doesn't sound like this statement can be right. The weather might be changeable but at any one instant of time, in any one place, experience tells us there is either a hurricane or not; there is no obvious meaning to the idea that both are happening. Yet the problem only gets more dramatic. In space, microscopic differences in the precise structure of gas clouds can, given chaos, make all the difference between a new star and planets being born, or the cloud evaporating ignominiously back into the cosmos. Following the same logic, quantum mechanics coupled to chaos appears to bring uncertainty not just to the weather but to the existence of entire planets, stars, and galaxies.

Everett was quite comfortable with such a conclusion, but intuition and common sense suggest that something crucial is missing from the account. The quantum pioneers, not least John von Neumann (whom I previously introduced as a nuclear-bomb developer and weather simulator), strongly believed that quantum fuzziness was a fundamentally small-scale phenomenon. In the face of chaos's tendency to amplify small differences, they invented special mechanisms to rein in the fuzz and keep it confined.

The central tenet of their traditional quantum mechanics is that fuzziness is switchable: sometimes it is on, and other times it must be off. A tiny electron, for example, spreads in a smeary haze for much of the time, but if you take a photograph with sufficient sensitivity, it will appear in a single, pinpointed location. While not quite a photograph in the traditional sense, machines that register the presence of a single electron really do exist, and they really do show just a single dot.[17] Yet a raft of other experiments, not to mention the entirety of chemistry, only make sense if electrons spend most of their time being fuzzy. So there must be a transition from fuzzy to definite, and it is known as *wave function collapse*. Von Neumann supposed this collapse took place "as soon as . . . a measurement is made."[18] After the apparatus stops measuring, the electron gradually resumes its fuzz, starting from its pinpointed location and spreading outward.[19]

It is not clear how to extrapolate from this traditional view of quantum mechanics to the case of weather. Presumably, the formation of a hurricane undergoes wave function collapse so that it has a clear outcome, but when, why, and how are all left unspecified— must some apparatus be measuring the storm development? What did von Neumann really mean by "measurement"? How does any particle decide when to switch from fuzzy to definite? These vexed questions have led to some of the boldest speculations in physics—things that sound to me like mysticism, put forward with perfect seriousness by leading researchers.

For example, Eugene Wigner, a Nobel Prize–winning physicist and mathematician, believed that quantum physics demonstrated a special role for consciousness.[20] His suggestion was that there is a concrete reality only because we conscious creatures are involved in measuring it, an extreme form of the philosophical school known as *idealism*. The notion that our *experience* of reality is inseparable from our minds is a less stark form of idealism with which I have no problem, but Wigner contended that reality *itself* is subservient to the mind in a way that I find hard to take seriously.[21] Everett's research supervisor, John Wheeler, was ambivalent about Wigner's special status for consciousness but nonetheless insisted that the past history of the entire universe is retrospectively determined by the observations that humans choose to make. "Equipment operating in the here and now," he wrote, "has an undeniable part in bringing about that which appears to have happened."[22] These were clever people, and one should try to understand their point of view, but the overall propositions seem sketchy and anthropocentric.

More soberly, the mathematical physicist Sir Roger Penrose has pointed to gravity as a possible physical mechanism by which quantum fuzziness is eliminated when dealing with sufficiently large objects.[23] Penrose's proposition and others along the same lines are far less mystifying, and attempts are underway to test them in laboratories.[24] Yet what we do already know from experiments is that the apparent speed at which the collapse process sweeps through space, cleaning the fuzz away, is faster than light.[25] This cuts against the grain of relativity in which light sets an absolute limit for speed.

In short, whatever way one tries to account for wave function collapse, the consequences do not sit comfortably with physics as we know it. That led Wheeler's student Everett to wonder whether collapse is real at all. The genius of his proposal was to reconcile the fuzziness and the certainty: he suggested that the experience of anyone living within a fuzzy world will seem just as concrete as ours does. Collapse will appear to occur, even if in reality it actually does not.

To achieve this magic, Everett recast fuzziness as a series of alternate universes superimposed upon each other.[26] From within each universe, things appear certain; yet, taken together, the universes are a *multiverse*, presenting a menu of possibilities rather than a single definitive reality. To put it more precisely, individual concrete universes such as the one we seem to inhabit give only a partial view of reality; they are like a shadow of the more fundamental quantum mush.[27]

One of quantum computing's pioneers, David Deutsch, has pointed out that in Everett's picture, quantum computers possess their extraordinary power for a clear reason: it is because their fuzzy symbols are making use of Everett's superimposed universes to carry out multiple calculations simultaneously, while a traditional computer relies only on one universe at a time.[28] If Everett is correct, your regular laptop exists in multiple universes, but it has no means of communicating between those universes. Like our own perception of a single reality, the machine is unable to access any hint of what is going on in the parallel worlds. This is not pure supposition; there is a mathematically demonstrable effect known as *decoherence*, which makes it exceptionally difficult to access information about the alternative realities even if they are present. A quantum computer, on the other hand, is engineered to avoid decoherence and so to make cunning use of a number of parallel worlds together. From Everett and Deutsch's standpoint, it is the need to maintain communication across these worlds that makes building a quantum computer so challenging.

Even if it makes logical sense and has high-profile backers, Everett's proposition, that the fuzz of quantum mechanics corresponds to a whole multiverse of possible realities, is deeply unnerving. It seems extraordinarily wasteful of universes. Isn't one enough?

Maybe not: I think we should mistrust human instinct on these matters. In the sixteenth century, there was zealous attachment to the idea that Earth is at the center of the universe; in the early twentieth century, eminent astronomers went to great lengths arguing that galaxies beyond our own did not exist. My suspicion is that many of the arguments against the quantum multiverse are just another example of the recurrent human trait: denying our own exasperating insignificance.[29]

Quantum Cosmology

Quantum physics describes a reality utterly different from the everyday world we inhabit. Individual microscopic objects are smeared into a mushy haze most of the time, but can become pin-sharp if anybody chooses to watch. Everett's way to make sense of it involves reimagining the mush as incalculably many parallel universes, all except one of which we will never experience. Despite seeming extravagant, I have argued that Everett's picture is better than the alternatives which involve new physical laws or even a special role for consciousness in determining objective reality.

Everett's theory implies that quantum effects are not after all limited to small scales, but stretch across the whole universe and beyond, if only one knows how to look for them. This kind of thinking emboldened physicists to apply quantum laws to the cosmic whole. In fact, it is key to making sense of simulations.

Recall from the previous chapter that the universe is expanding and, at some point in the past, had zero size—the Big Bang. Our simulations can't start with the Big Bang itself, because it has an infinite density, an infinite pressure, and an infinite expansion rate. The equations of general relativity break down when faced with these competing infinities; the result is known as a *singularity*. Just as with simulated black holes, where exotic wormholes were used to prevent the central singularities from appearing, the Big Bang singularity must somehow be avoided. It's not quite as complicated as with black holes, though; to avoid the Big Bang, cosmologists can start the whole simulation sometime after the initial moments of the universe. Typically, we choose around 0.1 percent of its 13.8 billion years.

The downside to this solution is that we need initial conditions representing the state of the universe at the chosen post–Big Bang starting time, in just the same way as weather-forecast simulations need accurate measurements of the atmosphere now to predict tomorrow's state of affairs. If the universe really emerged from a singularity, there is, by definition, no law to tell us what the initial conditions would look like. However, being the opposite of predictable, most cosmologists guess it would be wildly erratic.[30] Some regions might be cold and desolate, others hot and dense. One couldn't even take for granted that the laws of physics would be the same in all places. More likely, some regions would enjoy familiar physics, but others would be governed by riotously different laws.

This is far from what appears to be true of our universe, which is best described as samey. No single cosmic neighborhood seems differentiated from any other: planets, stars, and galaxies are, as far as anyone can tell, similar everywhere. That is not to say that all galaxies are identical. They vary enormously in size, color, and shape. But they do all appear to follow the same laws of physics, are composed from the same combination of gases and dark matter, and, moreover, the mix of galaxy colors, sizes, and shapes doesn't vary from one region to another.

It's like a fruitcake: on close inspection, some slices might have slightly more raisins and others more cherries, but the overall consistency is even. Following a Big Bang singularity, there is no obvious mechanism via which this regularity could be achieved. You'd be more likely to end up with a heap of raisins in one slice, a pile of apricots in the next, and some unexpected scrambled egg in the third.

Quantum mechanics is our best hope of finding principled initial conditions for simulations, and certainly provides new angles on the situation. First, it shows us that the equations were always destined to break; general relativity has no conception of uncertainty and entanglement that physicists know are crucial. If only we could include the quantum mechanical effects correctly, the singularity might be replaced by something more palatable. An attempt in this direction is the Hartle–Hawking *no-boundary proposal* made famous by Stephen Hawking's book *A Brief History of Time*. To Hawking and his collaborator James Hartle, a singularity was like a hard boundary to time itself; quantum mechanics would round that edge off, leaving something with no defined starting point and therefore no singularity.

But the proposal remains just that: a proposal, with implications that are still under debate. In practice, it has not so far provided us with much useful input into observational or computational cosmology, partly because we don't have a self-consistent description of quantum gravity—that is, the union between Einstein's general relativity theory of gravity and quantum mechanics.

These theories are fundamentally hard to combine. For example, the existence of black holes seems partially to contradict quantum theory: the former swallow particles and the information they carry about the universe which produced them, whereas the latter is adamant that information can never be lost in this way. Attempts to circumvent these difficulties and provide a workable description of quantum gravity have been underway for decades, and have resulted in an explosion of theoretical ideas including string theory, quantum

loop gravity, and causal set theory, among others. There is no shortage of ideas, but little in the way of concrete results that feed into cosmology.

Luckily, a second quantum angle on the early universe has provided much more in the way of testable predictions, again with crucial insight from Hawking. Instead of the no-boundary strategy of
replacing the singularity entirely—something which really depends
on physics that nobody yet understands—the alternative attempt
suggests that our regular universe will arise regardless of what happened in the opening moments.

While it draws on elements of quantum theory and general relativity, it doesn't require them to be fully interwoven; the brilliance is to
apply the theories to distinct, mutually exclusive aspects of the calculation, without making strong assumptions about how they should
ultimately be combined. I am going to explain the calculation in some
detail, because it is our current best shot at understanding the appropriate initial conditions for simulations, and it hints that quantum
effects suffuse the entire structure of the universe.

Inflation

In 1980, the cosmologist Alan Guth was thinking about the way matter and energy change as the universe ages. Ice can't exist for long
at room temperature; it turns into water. Water can't exist for long if
it's boiled; it turns into steam. But Guth knew that the phases of matter extend far beyond these everyday states. The theoretical physicist
Steven Weinberg had already proposed that even subatomic particles
like electrons, neutrinos, and photons cease to exist at sufficiently
high temperatures, instead turning into a purer type of energy. Guth

suggested that at higher temperatures still, all remaining particles could lose their identity and turn into an abstract form known as a *scalar field condensate*, a phenomenon only possible in quantum physics. If that were correct, he calculated that after the Big Bang, the universe would enter a phase in which it doubled in scale approximately every 10^{-35} seconds, driven by this strange form of energy. Such behavior is known as *exponential expansion*.

Scalar fields are not entirely hypothetical; their existence was indirectly confirmed by the Large Hadron Collider's detection of the Higgs boson, a particle which is associated with a scalar field, in 2012. But whether there is a scalar field that behaves in just the right way to power the early expansion of the universe is an open question. Alan Guth supposed there might have been, and traced the consequences.

In person, Guth is avuncular and gently wry, and his name for the proposed early period reflects that: he called it *inflation*. It's a nod to the other great exponential in our lives, the increasing cost of living. According to the US Bureau of Labor Statistics, $20 in 2022 bought only as much as $10 did in 1994.[31] That's a doubling time of twenty-eight years, although recently the pace has picked up to a far more problematic rate. In Germany, after the First World War, it was catastrophic: prices doubled twenty-nine times during the single year of 1923.[32] This conveys a little more of the drama of cosmic inflation.

Unlike financial inflation, however, rapid cosmic inflation is a good thing for physicists because it softens the problematic consequences of the initial singularity. For Guth's idea to work, the inflation must last for a minimum of ninety doublings, after which the rate of expansion decreases fantastically: today the universe doubles in scale only every 10 billion years or so, a much more leisurely pace.[33]

Because of this contrast, the most common description of inflation is as a short, sharp stretching of the universe early in its history, capable of straightening out any creases and folds, leaving a regular, uniform space.

Even if this description holds some intuitive satisfaction, it only captures part of the story. To understand the real power of inflation, it can be better to imagine cosmic history in reverse, turning expansion into contraction. In the backward story, during inflation the overall scale of the universe halves every 10^{-35} of a second or so. But half of something is never nothing. Cutting a piece of paper in half can't make it vanish; similarly, as we head back through inflation, space is ever smaller, but never zero size. By contrast, the backward story of a universe without inflation can reach zero size—the singularity—without any difficulty.

From this reverse perspective, inflation actually pushes the singularity very slightly further back into our past. Calculations that ignore inflation predict today's observable universe originally expanded from zero to the size of a football in less than 10^{-35} seconds. Calculations that include inflation multiply that time by around a hundred, since each halving takes the same length of time, and there are at least ninety of them. (The calculations are of course more complicated than this, and there is considerable uncertainty about the precise figures, but this gives the right outline sense of the effect.)

We are still talking about exceptionally short periods, but the increase from around 10^{-35} to at least 10^{-33} seconds nonetheless has major implications. Imagine a glassblower is creating a vase, starting from glass fragments with a riotous mixture of colors—a loose analogy for the singularity's expected disjointedness. In the normal Big Bang picture, the vase is blown in such a short time that the colors have no chance to mix: this is the unpredictably variable universe that I mentioned before, with each part of the vase being different from the others. But the hundredfold increase in time offered by inflation allows the molten colors to flow together, making the result much more uniform. The gently mottled glass is a much better match for the well-mixed cosmos that we seem to inhabit.

If the story of inflation ended there, it would be just a cooked-up explanation for what we already knew: space is pretty similar from one place to another. But there is more. Quantum mechanics forbids inflation from generating a universe that is perfectly the same everywhere. The uncertainty principle requires there to be small variations, with each small patch of the young universe containing either a tiny bit more or a tiny bit less material than its neighbors. To put it another way, our imaginary vase's colors will be well mixed, but some patchiness must remain.

As I've mentioned, if we look to greater distances in our cosmos, the light has taken increasingly long to reach us. With the right kind of telescope, it's possible to find radiation that is almost as old as the universe itself; this is a glow known as the *cosmic microwave background*. Variations within this light—patches of reduced and increased intensity—were measured from the 1990s onwards, and match beautifully with predictions from inflation made in 1982 by a number of physicists, including Stephen Hawking and Alan Guth.[34]

One way to visualize the scale of the predicted variations is to think of ripples on the surface of a calm ocean. The water is miles deep yet the waves atop it are a matter of an inch or two at most—barely noticeable. But if we include these kind of tiny waves in simulations with dark matter, gravity takes over; the ripples grow into galaxies and the vast, all-encompassing structure of the cosmic web around us today. Because the formation of stars and the solar system could only have taken place inside a preexisting galaxy, we conclude that everything we see, including everything here on Earth, probably owes its existence to random quantum effects in the first fraction of a second of the universe's existence. Quantum mechanics, gravity, dark matter, the cosmic microwave background, the web of galaxies, and our own existence: all are beautifully tied together in this vision.

The 1982 calculations did not dictate precisely what the ripples look like, which would contradict the uncertainty stipulation, but rather made predictions for the ripples' size and shape on average. It's the difference between predicting where every peak and trough of water lies on an ocean—clearly impossible—and making a statement about how many peaks and troughs can be expected, of what height, and at what separations from each other. Quantum inflation allows only these expectations to be calculated, providing us with a summary of the expected types of waves (known to experts as a *power spectrum*) but no detail of the particular waves in our universe.

And therein lies the headache for cosmic simulators. We went in search of the initial conditions for our universe. The hope was that, as for the weather, if we could input an accurate portrayal of the young universe, our simulations would predict what happened next: why we get a particular mix of galaxy types, what determines the characteristics of a particular galaxy, and how the Milky Way came to exist where it is. In short, we wanted a single, definitive history to place ourselves within a cosmic context. What we obtain instead is a random quantum froth, described by a summary power spectrum.

Cosmologists see this froth as specific ripples in the cosmic microwave background, and a specific set of galaxies around us here and now. But reality—if Hugh Everett is to be believed—isn't just one universe. It's an infinity of universes, containing all possible sets of ripples consistent with a power spectrum, leading to all possible consistent sets of galaxies, stars, and planets. The collection of universes each evolve according to the serendipity of their own random patterns. To put it another way, our specific universe was set in motion by the roll of multitudinous dice, and we have no way of knowing exactly which way they came up, so no way of perfectly re-creating our beginnings. We need to simulate the multiplicity of different possibilities before we can even begin to locate our own history.[35]

How are we to cope with simulating such a giant mishmash? Steady progress in quantum computing might one day revolutionize computational chemistry, but it is unlikely to ride to the rescue of cosmology simulators in the same way. The nature of our quantum problem is very different because the universe is overwhelmingly complex compared with a single molecule, so the power of quantum computers would have to be very much larger before they could contribute usefully. This level of sophistication may never be reached, and it certainly seems unlikely in my lifetime.

Meanwhile, each simulation on a traditional computer can choose only one of the possible universes to simulate, forgetting about the others. The individual virtual universe will not be like ours in detail, because the dice will have landed differently. That said, randomness doesn't imply complete unpredictability. This is familiar from everyday situations: for example, if you roll two dice and add their value, you will find that a total of twelve (which requires a double six) arises much less commonly than a total of seven (which can arise from several combinations, like six and one or five and two). It is legitimate to ask what trends and regularities emerge from random processes in the universe, even if an exact match is out of the question.

Many astronomers therefore focus not on the specifics of individual galaxies but instead the overall mix of sizes, shapes, colors, brightness, and so on. If we compute these for all galaxies over large regions of a single simulation, we can compare them with observations from similarly large regions of the real universe. It isn't just a case of checking one criterion at a time; the way that different properties correlate with each other (for example, the trend of size with number of stars, or shape with color) can also be checked. This verification process has been very successful in recent years; the overall mix of galaxies now agrees quite well with reality.[36] We are simulating something like climate rather than weather, verifying the overall patterns to be expected, rather than the specifics of our particular cosmos.

Success in this endeavor doesn't automatically translate into insight. If the purpose of a simulation is to interpret and understand the real universe, re-creating trends isn't in itself particularly valuable; what matters is to pinpoint the reason for the trends. In chapter 3, I gave an example of how simulations came to reproduce the shadows of dim, fragmentary galaxies in the early universe. This development would be relatively unremarkable if the simulation sub-grid had been specifically tweaked to accomplish such a match. It hadn't: the excitement was to discover how sub-grid processes designed to make sense of galaxies in the modern universe also make sense of their distant ancestors. The scientific value of studying trends in simulations is to make such connections, not to reproduce the trends for their own sake.

One can go a long way by studying statistical patterns, but there is a limit to what they can reveal. No galaxy is an average galaxy any more than any human is an average human. Worse, trends do not necessarily imply a direct connection ("Correlation does not imply causation," goes the aphorism). There are plenty of examples of this: for instance, people who shop in Neiman Marcus tend to be rich. That doesn't mean that being rich makes you shop in Neiman Marcus, and certainly not that shopping in Neiman Marcus makes you rich (if anything, the opposite might be true). Similarly, even if trends in the simulated galaxies agree with those in reality, it can be dangerous to jump to conclusions about why such connections exist. To understand what makes galaxies unique, another approach is needed.

Experiments with Simulations

In 2016, my colleagues Nina Roth, Hiranya Peiris, and I started wondering whether, despite quantum randomness, we could find ways to dig down a bit more into causal connections between the early and late universe—why things turn out the way they do.[37] Peiris is a longstanding collaborator of mine, with immense vision and a flair for asking big questions. Roth was a postdoctoral researcher in Peiris's group, and she had recently written a computer code that could generate suitable initial conditions for cosmological simulations. Like all such codes, it involved a random number generator—a way to create a specific universe from the multiplicity predicted by quantum mechanics. But, we thought, while inflation says that the early universe produces random outcomes, there is no need for simulations to follow suit.

Imagine you are interested in the possible outcomes of a game of snakes and ladders: instead of rolling dice, you could ask, "What would happen if I rolled a six? Or what if it were a five?" You might even try playing the scenario. It's not within the rules, but it does allow you to understand the range of outcomes.

Altering the initial conditions for a simulation so that they are no longer fully random is like performing this kind of what-if experiment within a virtual universe. Instead of accepting chance, Roth adapted her code to manipulate the quantum statistics in the early universe, crafting a series of alternatives. Even after we had simulated our galaxies once, we asked: What would they look like in a universe where the ripples were slightly different? How does cosmic history play out in alternative scenarios?

The name we give to this approach is *genetic modification*. It's a reference to how biologists can take genes from one species and splice them into another's DNA, studying the resulting organism. Similarly, we can edit the early universe and then perform the simulations again to study how the modified galaxies grow, comparing to the original version to understand the changes. The laws of quantum mechanics offer no way to change random outcomes in reality, but within the simulations we are free to try different possibilities. With an expanding collaboration, we've been able to isolate, for example, the multifarious factors controlling the brightness of galaxies and why some stop forming stars.[38]

We can go even further and make radical changes which turn giant clusters of galaxies into empty voids, by taking what used to be the crest of a quantum wave and turning it into a trough instead.[39] Manipulations like this give a fresh perspective on why some areas of the universe are remarkably empty, and in doing so increase the accuracy with which we can derive implications from observations of the cosmic web. This added accuracy may be crucial for understanding dark matter and its even stranger counterpart, dark energy.[40]

Dark energy is very weak, making it negligible in our own planet, solar system, and even galaxy, yet is found everywhere, including in the near-empty voids. This is why its effects can accumulate dramatically over distance, and in total it seems to outweigh matter; it makes up around 70 percent of everything in the universe. Moreover, matter (be it dark or visible) is diluted as the universe expands but dark energy is not; consequently, in the end, the latter will come to constitute close to 100 percent of all there is. Based on current extrapolations, over the next 100 billion years or more, dark energy will become sufficiently dominant to halt galaxy formation altogether, and what stars remain will fade and die. At this point, the scale of the universe will enter a pattern of doubling every 12 billion years, ever-expanding in a way that is strikingly reminiscent of inflation, albeit vastly slower. Just as inflation dictates the start of our universe, the presence of a tiny but ubiquitous trace of dark energy may dictate the end.

We cannot yet be sure. From 2009 to 2011, I shared an office with the astrophysicist and author Katie Mack, who—in spite of being a great office mate—seemed forever obsessed with cheerful discussions of the eventual demise of civilization. Then she literally wrote the book on them, so that we can all study varying versions of our ultimate demise.[41]

While thought-provoking, these very long-term effects don't cost me any more sleep than the destructive power of the black hole at the center of our galaxy. Dark energy is more immediately relevant as a clue to a deeper understanding of nature; perhaps it has something to say, for example, about quantum gravity. Experimenting with possible types of dark energy inside simulations, and understanding how quantum randomness from inflation interplays with gravity, dark matter, and dark energy to sculpt what we see today, is one way we have to make progress. As it stands, we're very far away from understanding what it is trying to tell us, but we have a digital laboratory where we can tinker away until we understand a bit more.

Did Inflation Really Happen?

To the pioneers—de Broglie, von Neumann, Bohr, and Heisenberg among them—the idea of applying quantum theory to the entire universe would be anathema. But oddball Hugh Everett showed there is no need for an artificial separation between small-scale quantum and large-scale cosmic phenomena. The two can sit quite comfortably together, provided we accept that our entire universe is a poor shadow of a more fundamental reality, one of truly terrifying scale. Building on this insight, and on the idea that scalar fields can power exponential expansion, physicists crafted a theory of inflation that explains the uniformity of our universe while also providing a mechanism for seeding the variety of galaxies that we find in reality. It explains the fact that our cosmic cake is so well mixed, and also accounts for its particular ingredients.

These ideas are a huge extrapolation away from laboratory-tested physics, and by no means all cosmologists are convinced that inflation is a compelling theory.[42] However, at the very least it serves as a useful proxy while we search for a more complete picture of what happened in the early universe. By studying the way that structure in our universe compares with simulations based on inflation, dark matter, and dark energy, we might be able to refine all the speculations, or supplant them with better ideas that don't yet exist. Meanwhile there are more predictions to test: the race is on to find evidence for gravitational waves that should have been generated during inflation. If these are discovered, the case for inflation will look stronger.[43]

But the case may never be completely decided. Unlike dark matter, it's hard to imagine that terrestrial experiments could ever verify inflation directly in a laboratory; the energies are around a trillion times higher than those harnessed by the Large Hadron Collider. Even if one could build an experiment to re-create these conditions, it might not be advisable. When the LHC started operating in 2010, concerns were raised that it might generate a black hole that swallowed Earth or, more dramatically, end the universe as we know it by destabilizing particles, causing a change of phase like the ones hypothesized for the early universe by Weinberg and Guth. These scenarios were assessed seriously, but ultimately dismissed because LHC-energy collisions occur regularly throughout the universe without any adverse effects.[44] That kind of argument couldn't be made for an experiment that seriously attempted to replicate the conditions of cosmic inflation; such an experiment could indeed end the universe as we know it, which would be embarrassing. Just as well it is so firmly out of reach.

Even if many of the theories and phenomena I've described in the book so far are tentative, simulations of our universe have no choice but to include them: dark matter, dark energy, the sub-grid details of stars and black holes, and now inflation and its implications for cosmic initial conditions. The discussion of other phenomena—magnetic fields, for example, or little chunks of matter called *cosmic rays* hurtling through the cosmos at near the speed of light—could take up books of their own, and their effects in simulations are being studied intensively. As yet, however, they do not lead to major differences in our understanding but rather refinements to the detail. By its nature, a simulation will never be exhaustively complete, but I have outlined the most important ingredients for describing galaxies and their implications for the broader cosmos, at least as we currently understand them.

Having dealt with the ingredients, it's time to reexamine the results.

Simulations do not produce predictions that we can straightforwardly compare with what's really out there in space. All simulations are approximate; chaos magnifies even the slightest inaccuracy to cosmic scales; and inflation gives a huge range of possible starting points rather than a single clear description of the beginning of time.

All this means that simulation codes cannot capture anything beyond general guidelines for how the universe works, in the same way that climatologists can't tell you the precise weather a century from now. And yet cosmologists still hope to compare the real universe with simulations, to infer something about the nature of dark matter, or the rate at which dark energy is pushing our universe apart, or the physics that controlled the origin of everything, 13.8 billion years ago.

To accomplish this requires intelligently sifting vast quantities of data that automated telescopes are gathering. That data must be compared with simulations, but not in any simple spot-the-difference way. Part of a cosmologist's job is to distinguish the wheat from the chaff: to decide what is in agreement between our computer worlds and the real one, what is truly different, what might be the result of random fluke, and what just isn't yet understood. No human can digest all the data held on the universe, nor all the results from simulations. As a result, we are increasingly leaning on computers to do the job for us, and that requires simulations of a completely different type: simulations of thinking.

6

THINKING

Machines with intelligence have long been a human dream. In Greek mythology, the god Hephaestus crafted artificial creatures that could move, interact, and think. His creations, according to Homer, included two metal guard dogs which remained "deathless forever and un-aging."[1] Very handy.

The mechanical prowess of an automaton that can replicate a dog's movements would be impressive. The process of synchronizing the motion of four limbs to move efficiently over uneven terrain requires adaptability: it is continuously necessary to scan the ground's three-dimensional contours, creating strategies for traversing it safely, then turning the abstract plan into specific physical leg movements. On top of that, to serve as watchdogs, Hephaestus's robots must have been able to perceive potential threats around them and rapidly decide on a response.

That requires a level of intelligence and self-determination that we rarely associate with computers. Yet, in the early twenty-first century, humanity is finally able to achieve aspects of this sophistication in real-world automata—controversially, mechanical dogs have even been trialed by police forces.[2] When put to work, these and similar robots appear eerily lifelike. There is something about the style of the motion, combined with an obvious focus on a particular goal, that conveys personality. In one publicity video, a human attempts to prevent a robodog opening a door—the ensuing short story of the robodog's triumph over adversity delivers an emotional punch.[3]

Here, emotion is in the eye of the beholder: there is no suggestion that these robots are conscious. But they are undoubtedly intelligent. The distinction is crucial because the origin and meaning of consciousness is thorny, whereas intelligence can be made relatively simple: if something reliably behaves in an intelligent way, one can choose to accept that it is intelligent by definition. Humans, dogs, spiders, and slugs are all intelligent in their own ways, so evidently there are degrees to consider. But to avoid enormous digression, let's assume we recognize intelligence when we see it, and leave emotion and consciousness to the philosophers and neuroscientists for now.[4]

To create artificial intelligence in machines, coders somehow need to turn rigidly rule-following computers into flexible thinkers. That sounds absurdly ambitious, but previous chapters have shown how simulations of a dark-matter halo, a galaxy, a black hole, the universe, or even a single molecule are also absurdly ambitious. The trick with these physics problems has been to create a simplified virtual mimic of reality, later adding detail and nuance in incremental layers if required.

Intelligence can also be mimicked without precisely copying a human or animal brain; all that matters for any practical purpose is the behavior presented to the outside world. Alan Turing captured this insight in 1950, suggesting a test in which human interrogators engage in a conversation through text messages, without knowing whether the other party is human or machine.[5] According to Turing, computers should be regarded as intelligent if, after a prolonged conversation covering whatever topics are desired, the interrogators cannot reliably tell whether they are talking to a human or a machine.

The test is not perfect, and plenty of objections to it have been raised. Apart from anything else, it assumes a narrow vision of intelligence in terms of human language. What of art, sport, or music? This is a fair objection, but Turing's overarching point is deeper. He wished to highlight that inspecting a machine physically can't tell you whether it is intelligent any more than a physical examination of someone's brain tells you their aptitude for any particular task. Our only option is to assess behavior.

In the early 2020s, simulations of thinking seem to be developing toward assured intelligence at great pace. As with any significant technological development (think of printing, the spinning jenny, steam engines, electricity, fertilizer, cars, or the internet), the disruption to society is likely to be profound and hard to predict. At the moment, machine intelligence is removed from human levels in terms of its overall adaptability, but, faced with certain narrow tasks, it has become sufficiently accurate and flexible to take on skilled jobs: keeping records, searching for information, identifying faces in images, driving trains, interpreting medical scans, predicting shopping habits, writing simple essays, and even performing basic legal analyses.[6]

Artificial intelligence is also becoming an unavoidable part of many sciences, including cosmology. To see why, consider the Vera Rubin Observatory. The telescope isn't used to zoom in on individual objects; instead, it is conducting a fully automated survey of the sky, scanning to see what is out there. From its vantage point atop a Chilean mountain, the telescope is expected to discover, classify, and monitor around 20 billion galaxies. At the same time, it will find and determine the trajectories of thousands of asteroids, checking whether any pose a collision risk (not so different from Hephaestus's robots scanning the horizon for threats). The raw information resulting from this exercise consists of fifteen terabytes—the equivalent of ninety cinema-quality movies[7]—arriving each night for ten years.

Turning the raw outputs from a telescope like this into information about our universe requires extensive data processing. One has to identify the separate objects in each individual photograph, classify whether they are stars, galaxies, quasars, asteroids, or something else, and compare against previous exposures to determine whether the object is moving or has changed. Cosmologists will then want to estimate the distance to supernovae, galaxies, and quasars to construct a three-dimensional map of the universe. Finally, the map can be compared with cosmological simulations to determine its implications for physics—whether it tells us anything new about dark matter or dark energy, for example. All this is far too much to be attempted by humans alone.

For at least two decades, astronomers have been working with artificial-intelligence techniques to automate each step of the process. One way to begin is by building a digital brain, loosely inspired by the physical functioning of our own brain's neurons. The resulting computer code, like a baby, has no intrinsic knowledge or abilities: it must be trained to perform the required tasks. We show the fledgling thousands or millions of instructional examples—these might illustrate the difference between known stars, quasars, and galaxies, or more generally will exemplify whatever work we wish the machine to undertake. The simulated brain responds by changing the connections between its virtual neurons in a way analogous to biological learning. After that, it is ready to work.

This approach, known as *machine learning*, has immense flexibility. But it can be exceptionally hard to understand what the computer has learned, why it reaches particular conclusions, and whether one can rely on those conclusions for scientific deductions.

I will start instead with a complementary vision of intelligence that has the opposite characteristics: it is rigid but transparent and rigorous in its thinking. This approach is based on *Bayesian statistics*, a small set of logical principles that describe an idealization of rational, scientific thought. Instead of a loose analogy to a malleable biological structure, the computer code contains precise descriptions of allowed steps in the reasoning, alongside a description of what we already know—how stars, quasars, and galaxies shine, for example. One great advantage of this method is that it allows existing human expertise to be coded directly into the computer, unlike a machine-learning approach where everything must be learned afresh.

Life on Mars

One of the first to recognize the value of a logical, pre-educated form of machine intelligence was the prominent chemist Joshua Lederberg. In 1942, aged just seventeen, he was already well into his undergraduate studies at Columbia University, and started to work part time in one of their research laboratories. Under the tutelage of the lab's director, Francis Ryan, Lederberg developed a fascination with organic chemistry and life. Ryan's wife recalled, "You could tell that Joshua was in the lab because you could hear the tinkle of breaking glass. . . . His mind was far ahead of his hands."[8]

By 1960, Lederberg was a Nobel laureate, a pioneer of molecular biology, and a consultant to the US space program. NASA was in the process of planning the Viking missions, an ambitious project to land on Mars and search for signs of life, and Lederberg helped design the instrumentation that would determine the chemical composition of the Martian soil.[9] The detector—known as a *mass spectrometer*—was similar to a modern drugs-and-explosives detector at an airport.

These devices take a highly indirect approach to detecting and classifying molecules. There is no question of developing a microscope that can zoom in and take pictures of the molecules in a sample; the constituents are far too small and numerous for that. Instead, molecules are destroyed by bombardment with electrons, and the spectrometer measures the masses[10] of fragments in the sample, forming a unique but highly cryptic thumbprint for the chemical under study.

Prohibited compounds can be passed through one of these mass spectrometers, with the results stored for later reference. In fact, if one knows the chemical structure of a substance, a computer simulation can reproduce the *bombardment effect* and predict how it would show up, eliminating the need to get any hands dirty in a lab.[11] Either way, given a new thumbprint taken from an unknown test substance, a human or computer can search through the accumulated library of known compounds, checking for a match.[12]

To scientists, this is known as solving an *inverse problem*, and while it is possible in principle it can be hard in practice. If you have a criminal in front of you, it's quick and easy to take a record of their fingerprint. But the inverse problem is difficult: you have a crime-scene print, and you need to identify the criminal, involving a tedious search through records. Worse than that, if you are searching for signs of life present or past on Mars, you have only a faint idea of the kinds of organic molecules you might encounter—quite possibly things that do not exist on Earth. Lederberg realized that the Viking mission needed a system to infer the structure of new molecules. It's like solving a whodunit on another planet, using prints from a criminal you may never have even encountered before.

Given enough time, humans might approach this inverse problem along the following lines: have a guess at the possible chemical structure, simulate that structure's thumbprint, then compare the prediction to what has actually been recorded on Mars. If this doesn't result in a match, start again with a new guess. The approach is sound, but can be exceptionally tedious because there are so many reasonable guesses for possible molecules.

Lederberg saw that a computer could replace human intelligence in this search, automatically making suggestions for possible structures and comparing their signatures with reality. In 1965, while working at Stanford University, he met Edward Feigenbaum from the Department of Computer Science. Feigenbaum was already interested in replicating scientific thought processes within a computer, and Lederberg provided him with the perfect problem to work on. The result was a project that lasted two decades, and that has become legendary in artificial-intelligence circles: DENDRAL.*

DENDRAL's chain of logic progressed through several key stages. First, it listed all possible compounds involving a known set of elements. Next, it looked at the mass spectrometer results and guessed— using large numbers of preprogrammed rules contributed by human

---

* The name derives from *dendritic algorithm*. *Dendritic* refers to the tree-like branching structure of organic molecules.

experts—which of these compounds were plausible. Finally, it used a physics simulation to generate a detailed thumbprint for each plausible candidate and compared these with the reality. Overall, the result was to jump quickly from the abstract mass spectrometer results to a concrete chemical structure.

The approach worked beautifully, but the results were disappointing: neither the Viking landers, nor any craft since, has detected conclusive evidence that life has ever been present. The search isn't over, though: at some point in the late 2020s, it will continue when ESA's *Rosalind Franklin* rover, which carries a mass spectrometer, lands on Mars.[13] *Franklin* will drive over the surface, stopping in different locations to drill two meters down, looking for evidence of life deep underground. If any tentative signs are discovered, inferring their exact chemical composition will become one of the most important scientific projects of the age.

Bayesian Logic and Spinning Universes

Inverse problems similar to this are ubiquitous in science, including in astronomy: reconstructing the star formation history of a galaxy, determining the atmosphere of distant planets, weighing the contents of the universe, or searching for supernovae would once have been a task for human experts sifting through images. Following a similar logic to DENDRAL, we can replace their labor with repetitive, mechanical machine manipulations. The computer lists all candidate explanations for a given observation, calculates what would have been seen assuming each of the individual possibilities, then compares with reality to establish which explanation is the best available.

But the approach so far has a key missing ingredient: uncertainty.

All our knowledge is fuzzy to a degree. At this moment, I have no idea whether there was ever life on Mars; if a future space mission finds tentative evidence, it might increase my confidence without making me certain. Similarly, I am not completely sure whether dark matter exists, but given the evidence and the lack of plausible alternatives, I think it's pretty likely.

Even when uncertainty seems to have been eliminated, part of the job of a scientist is to remain cautious because we have incomplete understanding of our experiments and instruments, which anyway are only ever capable of making less-than-perfect measurements. Consider the case of dark energy. At the time of writing, the most precise methods suggest that 68.5 percent of the universe's contents are in the form of dark energy, but it could easily be another figure within 2 percent of this estimate.[14] The uncertainty has shrunk over time, as improved technologies have allowed for more accurate measurements. And yet even the *uncertainty* must be viewed with caution: if we allow for the possibility that general relativity is not entirely correct, it is conceivable that the phenomena we attribute to dark energy reflect our incomplete understanding of gravity rather than the presence of any energy at all.[15] So, doubt can arise from unavoidable imperfections in our measurements, and from the tentativeness of the scientific theories involved.

To take an extreme example, there's no serious question in my mind that the sun will always rise in the east. But one ought to permit some residual doubt to allow for the remote possibility of a law of physics that humanity hasn't discovered yet, dictating that the direction of Earth's rotation will suddenly reverse next Tuesday night, causing the sun to rise in the west on Wednesday. It seems vastly unlikely, but hard to rule out on rational grounds alone. Philosophers of science call this the *problem of induction*—no amount of past experience would seem logically to rule out a change in the future.

Worries like this are more esoteric than practical, but they do illustrate that whenever we quote even a simple scientific result, it carries with it a range of doubts. Sometimes, as in the case of the rising sun, those doubts can be left implicit; other times, as in the case of dark energy's quantification, it is paramount to include them. If a computer is to reproduce good scientific reasoning, it will need to follow the kind of logical, methodical approach of DENDRAL while allowing for these layered notions of doubt. Luckily, a perfect framework for that allowance already exists: Bayesian statistics.

Instead of associating each proposition with truth or falsehood, scientists can associate it with a number between one and zero; these numbers are called *probabilities*. If the number is zero, the proposition is claimed to be undoubtedly false. If it is one, the statement is supposedly inarguable and 100 percent true. But statements about the real world have probabilities in between these extremes, since nobody can ever be certain. If I find evidence in support of an idea, the probability for that idea should move closer to one; if the evidence seems to contradict the idea, the probability should move closer to zero. A flexible robot scientist must similarly be capable of reasoning in the presence of shades of gray.

Suppose I arrive at a cafe at one o'clock and place an order. Based on my past experiences, I have a fair degree of confidence that some food will arrive within half an hour. In Bayesian terms, the probability for the food arriving is pretty close to one. But as time goes on, nothing comes and I begin to wonder whether the order may have been forgotten: the probability of it arriving within the time window declines. I glance around, spotting other customers all waiting and looking at their watches. The probability declines further. Unable to gain the attention of any staff, with seconds to go before the one thirty deadline, the probability drops close to zero. But then, at the last moment, the meal arrives! Suddenly, the probability zooms back up to one.

The cafe scenario illustrates how probabilities reflect degrees of belief, and are accordingly malleable as new information arrives. They can swing dramatically from one moment to the next, and they may differ radically between people: the chef preparing the food, or the waiter who has seen that the kitchen is overstretched, may individually hold very different probabilities from mine. Despite being different, none of these probabilities is incorrect; rather, they are all conditional, and the differences reflect the various knowledge which the characters possess.

So far, I have only discussed the reasons for probabilities increasing or decreasing, leaving open the question of precisely how much the numbers change. In the case of a cafe, the numerical probabilities are rarely quantified or compared, so may not seem to matter. In the case of a robot scientist that uses probabilities to assess new information about our cosmos, it is essential to know whether a given piece of evidence would shift the balance of probabilities decisively, a little bit, or hardly at all. The central claim of Bayesian statistics is that there is only one sensible way to update probabilities in the light of new information. This change is determined through an equation known as *Bayes' theorem*, explaining why the whole machinery is called *Bayesian probability*, *Bayesian logic*, or Bayesian statistics. (In fact, the contribution of the eighteenth-century clergyman Thomas Bayes to establishing the field of study that bears his name was rather peripheral—the physicist Pierre-Simon Laplace played a more central role—but Bayes's name remains firmly attached.)[16]

Bayes' theorem expresses in mathematical form the researcher's mindset that nothing is certain, that nobody can tell us decisively what to believe, but that new evidence should change our opinions in a predictable way. In astronomy, the practical importance of Bayesian probability can't be overstated. It has become crucial for distilling cosmic microwave background observations into estimates of the composition of the universe, for deciphering gravitational waves to understand black holes, for inferring the properties of distant planets, and for weighing the dark matter within the Milky Way.[17] These are all tricky problems for which no single piece of evidence is decisive; Bayes (or, more accurately, Laplace) gave us a framework which integrates all the different observations and results into a unified assessment of what is likely and unlikely.

Bayesian probability has been central to my own research, and one of my favorite examples of its application concerns the way the material in the universe is moving. In chapter 2, I mentioned that Vera Rubin had wondered whether the universe might have an overall rotation, incurring the wrath of the *Astrophysical Journal* editor and other established experts who told her the question was misplaced.[18] Today we have a different perspective; the question is actually vital. Theoretical calculations in the 1960s and 70s by Stephen Hawking pointed out that it would be entirely possible for the universe to have a spiraling motion, but if something like inflation occurred, the spiraling would be stamped out.[19] Determining whether or not there is any overall rotation therefore helps us understand more about the very early universe.

A few years ago, I applied Bayesian logic to this problem. After Rubin stopped searching, Hawking had performed some preliminary calculations of how the light from the Big Bang would be distorted in spinning scenarios;[20] during my own PhD I took time away from simulations to perform a more complete calculation of the effects.[21] The prediction is for psychedelic swirls in the cosmic microwave background, as though someone has stirred hot and cold patches through the universe. While no such swirls were obvious in the data, their size and intensity would depend on exactly how fast the universe rotates, and a very slight residual spin could be partially obscured by quantum ripples.

In 2016, my collaborators and I worked with a new student, Daniela Saadeh, to comb the cosmic microwave background for the subtle swirls. It is not an exercise anyone could undertake without a computer. For a start, there is an overwhelming amount of information (comparable to an image from a fifty-megapixel digital camera). More than that, this gargantuan image must be compared not just to a single signature but to an overwhelming catalog of possibilities—a spinning universe could spin at any rate, in any direction. Because of all these factors, humanity will never have a definite answer to whether the universe spins, but Bayesian probability allowed us to calculate the odds.

The probability for the universe spinning was an astoundingly small 0.0008 percent, or one part in 121,000. The computer code played the role of a diligent expert, telling us that after an exhaustive search, while it can't completely rule out a spinning universe, the possibility is exceptionally remote. We took this as another sign that the inflationary picture, in which spin is excised from the early universe, holds some water.

The rotating universe was a rather niche project, although it's nice to know that we are not in a cosmic tailspin. But there are many more mainstream uses of the Bayesian approach, not least to quantify the contents of our universe and its rate of expansion. Among these, one particular example bears study because it illustrates how astronomers have come to rely on machines to perform routine but crucial steps in pretty much every analysis—and why that sometimes poses serious problems.

Redshift

Observational astronomy can be divided roughly into two activities: studying individual objects and generating maps of where these objects are strewn through the universe. Geographical maps are familiar to us, but some of the oldest known paintings—such as those created 16,500 years ago inside the Lascaux caves in France—may instead chart stars in the sky.[22] Today, mapping space reveals the cosmic web and has the potential to tell us about dark matter and dark energy. Yet to achieve this potential, astronomers must first accept that the position of stars and galaxies as seen in a two-dimensional image tells only part of the story. To understand the universe, we need to incorporate a third dimension, depth.

By far the most common way to add depth to these two-dimensional images is using an effect known as *redshift*: as light travels through the universe, its color becomes progressively shifted. A beam of blue light, for example, will look green after it has traveled for a few billion years; given a few more billion years it becomes red, and then continues shifting into the invisible infrared. The underlying cause is the expansion of the universe—you can imagine the light being stretched, which causes the change in color. The farther away the galaxy, the longer it travels for, and the more its light is therefore stretched and redshifted. If astronomers measure the apparent color of a particular galaxy, they can gauge how far away it lies.

To use this effect to create three-dimensional maps, astronomers need to know the original color of the galaxy; otherwise, they cannot tell whether the light has reddened over time or was red to start with. Stars look close to white, but if you allow your eyes to adjust to the dark night sky, you will begin to see a rainbow of colors, albeit muted. Take a look at the Orion constellation and you find Rigel, almost blue, next to Betelgeuse, distinctly red. In the cases of these relatively nearby stars, you can be sure that the colors are intrinsic, and nothing to do with distance or the expansion of the universe. But when astronomers discover much fainter pinpricks of light in the night sky, they cannot at first be sure whether the color is intrinsic or an artifact of redshift.

To explain how artificial intelligences can disentangle the original color from redshift, I need to dive a little further into the physics of light and vision. What we perceive in the night sky as pastel off-white colors is, in fact, a complex soup of information. If you look at bright stars on a dark night, each is providing several hundred thousand *photons*—little parcels of energy—to your retina every second. Each individual photon carries its own specific color, but the combined action of your eye and brain takes the resulting hundreds of thousands of colors and turns them into one.

Eyes characterize the color of light by estimating the number of photons that are reddish, the number that are greenish, and the number that are blueish. A yellow photon, being intermediate between red and green, can count in both the red and the green categories; similarly, turquoise light activates the green and the blue receptors. Your brain reassembles a single perception of color from this information. It's a bit like describing a demographic by giving the number of children, adults, and pensioners: it gives you a rough idea of the population's profile, but it's a far cry from knowing the distribution of ages with precision.

Our restricted color vision is rather pitiful; if there were a way to experience the reality of color, it would be vastly richer.* Measuring something closer to true color can lift the confusion between distance and intrinsic redness. Before it starts traveling through the cosmos, light emitted by stars, galaxies, and quasars is already a lavish combination of countless colors in a variety of proportions. Some mixes are more common than others, and some mixes cannot be produced naturally—except by redshifting. Provided the telescope is able to record colors in a more complete way than human vision, we have a good starting point for estimating redshift.

A separation of color into its components is known as a *spectrum*, and I previously discussed how it can help us with understanding dark matter and galaxies. A human expert can quickly inspect a spectrum and tell you a vast amount of information: Is it a galaxy, a star, or a quasar? If it's a galaxy, does it contain mainly new or old stars? And, crucially, how much has the light been redshifted? These spectra act as fingerprints for stars, galaxies, their contents and distance in much the same way mass spectrometers generate fingerprints for individual chemical elements. Working from the fingerprint back to the degree of redshift is an inverse problem.

Solving this particular inverse problem with computers is essential

---

* That said, our limited perception can also be helpful, at least for the entertainment industry: when watching the television, our eyes are convinced they are seeing a vast variety of colors when in fact the screen simply mixes red, blue, and green light in different proportions. To an objective observer, the colors generated by the screen would have very little to do with the colors of the true world, but to our vision systems, the illusion is utterly convincing.

for cosmology: there are insufficient experts in the world, with insufficient patience, to process billions of galaxies that are being observed by the Vera Rubin Observatory. Human inspection is impossible at this scale, and even taking detailed spectra is a distant dream because it takes a telescope too long to split the light of each individual galaxy into a multitude of colors. Instead, the observatory measures the intensity of light for a handful of different colors, an approach somewhere in between the limited abilities of the human vision system and the inaccessible extravagance of taking spectra.

The machine compares pictures taken in this way with colors predicted for all conceivable types of galaxy—the different sizes, ages, and chemical compositions, as well as the possible redshifts. Typically, there is not a single clear-cut answer for the redshift and the computer can only provide different probabilities, rather than picking out a single winner.[23] The result is a map of the universe that is smeared in the third dimension, reflecting the level of doubt in the computer's thinking. Such fuzzy maps are crucial for 2020s cosmology.

Known and Unknown Unknowns

Smeary redshifts constitute just one of many sources of uncertainty that cosmologists encounter when drawing conclusions about the expansion rate of the universe, the quantity of dark matter, or the strength of dark energy. Other sources include the quantum randomness from inflation, the limited number of galaxies, and more down-to-Earth problems like imperfections in telescopes.

Bayes' theorem is comfortable with this situation. Suitably programmed computers are able to combine these tributaries of uncertainty into a river of doubt and draw correspondingly reserved conclusions. But there is a serious limitation: this approach characterizes a flow of uncertainty but does not provide a guarantee against missing one of the tributaries altogether.

Some uncertainties we can at least sketch out. For example, we know that a red, dim galaxy nearby might be mistaken for a blue, bright one far away. Other uncertainties are far less definite, expressing the possibility that something as yet unidentified is missing from our account of a situation; these are the *unknown unknowns* made famous by Donald Rumsfeld. Our account of the way galaxies shine and change over time is incomplete because the intricate life cycles of stars within galaxies are exceptionally complex and varied, and when working with billions of galaxies, one should expect to see plenty of rare oddballs with the strangest properties we haven't yet dreamed about. It is hard to include these kind of unknown unknown possibilities in a Bayesian probability and yet, if we ignore them, the final uncertainty in our three-dimensional maps may be vastly underestimated.

Bayesian logic provides a useful and philosophically attractive account of scientific reasoning, but in practice is bound up in its initial conception of the world. This isn't a criticism. It's built into the framework: probabilities are able to carry very precise meanings because they refer back to a determined, preexisting body of knowledge. By contrast, humans are approximate but flexible thinkers and quick learners who can seamlessly adapt to the unexpected.

To see why biological brains are so impressive in this respect, imagine a game of tennis. The ball is traveling toward you at speed, and you have to decide how to respond. Before you can, it's necessary to work out where the ball is headed. Before even that, you need to gauge where the ball is now, starting from a confusing field of vision packed with distractions. Your brain is able to process these different steps in a flash, using skills learned through trial, error, and repetition—from practicing tennis over many years, from playing with a ball since your youth, and from using your senses to understand the world since you were an infant. Such learning isn't of an academic, scientific, narrowly Bayesian type; it's innate, and in some ways superior.

Let's imagine I build you a robot opponent, which can hit the ball as deftly as any human. It has been programmed with the laws of projectile motion, has an impeccable vision system with laser-guided ranging, and understands all the rules of tennis. It can use Bayes' theorem to change its beliefs about which shots you are more likely to play and which mistakes you are more likely to make, adapting its strategy accordingly. In short, it seems ready to win every match.

Yet if my robot is rigidly Bayesian, you may still be able to defeat it. When using Bayesian probability, I must specify in advance what can be learned. Suppose I forgot to code the aerodynamic laws that describe how spin affects trajectories, and you choose to slice the ball; the robot will be foiled. Even more embarrassingly, if I didn't include a way for the robot to learn physics as well as strategy, it may never learn from its error. No matter how many times you defeat it, my robot will make the mistake time after time because spin is an unknown unknown. The ball might as well obey laws of physics from a different universe.

I could anticipate and fix the aerodynamics omission, giving the robot a preconception of spin or at least allowing it to learn. But what if you challenge my robot to play on a different surface? Or on a day when the wind is gusting? Again, unless I have prepared the robot to adapt to these specific factors, it won't be able to do so. Whatever holes in its programming I patch up, you can probably think of another factor which has still been omitted.

Humans are more inherently adaptable without having to be prepared for each specific uncertainty. They will quickly cotton on, even if they haven't been trained in aerodynamics or played on clay before. This adaptive facet of intelligence is very hard to reproduce within the formal framework that I have sketched so far. There is nothing wrong with Bayesian probability, and in the right circumstances it is exceptionally powerful. But when unknown unknowns threaten, it makes sense to lay aside the idealized vision of perfect scientific reasoning, and instead study the way that our human brain achieves imperfect, flexible, creative thought.

Neurons

Our brain is built from *neurons*, which control electrical signals that carry and process information; loosely, they are the equivalent of transistors in computers. But while transistors come in only a handful of flavors, each providing a single switch that can turn an electrical signal on and off, neurons are diverse and multifunctional, monitoring thousands of inputs and combining them in varied and complex ways.

At its simplest, a neuron generates a pulse of electrical activity if, over a short space of time, it receives a sufficient number of incoming pulses from other neurons, or from one of our senses. Some inputs have a strong effect, easily activating the neuron; others are far weaker, only having an effect if accompanied by others. Inputs can even be wired to have a negative effect: a pulse arriving to these provides a temporary override, silencing the neuron's output regardless of how much it is encouraged to activate. And neurons can exhibit complex signatures far beyond this, such as firing electrical signals with a repetitive rhythm.[24]

The first simulations that captured some of this complex functionality, starting from the physics which governs movement of charged particles, were performed in the 1950s by Alan Hodgkin and Andrew Huxley; the pair duly won a Nobel Prize.[25] But while mimicking the biophysics is impressive, it is a long way from simulating thought. That is partly because of the sheer number of neurons required: your brain contains almost 100 billion. It has been estimated that simply imaging a human brain at the resolution necessary to map out the neurons and their connections would take around 2×10^{21} bytes[26]—a significant fraction of all the computer storage currently on planet Earth.[27]

Even if the profound challenge of snapshotting a single brain's wiring diagram could be conquered, it would not be enough, because brains change when they learn. Physiologist Ivan Pavlov is famous for having observed that dogs whose mealtimes were regularly accompanied by a background noise—something specific, like the beat of a metronome—would subsequently start salivating at the mere sound. In 1949, Donald Hebb proposed that this kind of association could have a physical basis at the cellular level, if two neurons firing repeatedly in sequence tend to strengthen their subsequent effects on each other.[28] At first, food and metronome concepts correspond to near-independent neural structures, but over time, any initially tenuous connections are incrementally fortified until there is a strong link.

Hebb was a psychologist and had investigated how people could learn to recover cognitive function after brain surgery.[29] His proposal was based on intuition built from these studies, rather than any understanding of neurons; modern experiments confirm the loose idea, but highlight that the exact way neural wiring changes over time is hard to predict in detail.[30] On top of this, the brain's electrical properties are modulated by hundreds of chemicals, the most important of which are associated with mood and feelings of pleasure, helping brains to learn through reward. No surprise, then, that even simple organisms remain a mystery: the entire nervous system of the tiny, one-millimeter-long nematode *Caenorhabditis elegans*—302 neurons and around 7,000 connections—was mapped as long ago as 1986, but there is still little prospect of simulating its behavior in a computer.[31]

Simulations of neurons are invaluable for understanding the brain's function, and so potentially to help develop life-saving medical insights. But for astronomers, scientists, and engineers, whose interest is to mimic the flexibility of human thinking in a computer system, it is unnecessary to understand every detail of the brain and literally re-create it in a digital mock-up. Instead, we use systems which are more loosely inspired by neuroscience, keeping the essence of flexible learning while ditching the complications of biology.

Machine Learning

In 1958, thirty-year-old Frank Rosenblatt wrote an article with a remarkable claim: he was building "a machine capable of perceiving, recognizing and identifying its surroundings."[32] Rosenblatt expected to spark a revolution in our understanding of intelligence; he compared the significance of his discoveries with those of Isaac Newton in physics, and told the *New York Times* it was possible in principle for his machine to become conscious.[33] (He would later distance himself from these claims, blaming them on "the popular press" which had "the sense of discretion of a pack of happy bloodhounds.")[34]

Rosenblatt was ambitious, persuasive, and possibly a little unhinged: after buying a $3,000 telescope to indulge his interest in astronomy, he realized he had nowhere to put it and persuaded some graduate students to build an observatory in his garden.[35] But his lasting accomplishment was a machine that he named the Perceptron, a device capable of learning to distinguish between letters, shapes, or other patterns presented to a camera. In itself, this wasn't especially impressive; what was new about the Perceptron was that it did not have to be programmed with code as we normally imagine it. Instead, it could learn through trial and error, as humans do.

The machine used a 20 × 20 grid of black-and-white receptors to turn an image into electrical signals, much like our own retina. The signals were then processed by a series of neurons—not real ones but electrical circuits designed to behave a little bit like them. The inputs to the first sixty-four neurons were wired from the receptors at random, with no designed pattern. That would never work with a traditional computer, but it turns out to be a good starting point for a machine that learns, analogous to an infant brain. The outputs from the first set of neurons were wired randomly into another set of neurons, and from these into two final neurons which connected directly to a pair of light bulbs. The Mark 1 Perceptron had more than a passing resemblance to a rat's nest: all these connections required a lot of crisscrossing wires.[36]

The goal was for the machine to distinguish pictures, one test case being whether it could learn the difference between simple shapes. The human operator showed the Perceptron squares and triangles over and over again. At first, it understandably responded by lighting its bulbs at random; but the machine had the power to change for itself the strength of the connections between different neurons. Inspired by Hebb's thinking, Rosenblatt arranged that, if one of the bulbs lit, the strength of the connections would automatically be tuned to discourage the other bulb from lighting, and vice versa. Over time this would naturally begin to divide the objects presented to the system into different classes; one bulb would indicate "triangle," while the other would indicate "square."[37]

At the time of his death in a boating accident (while he was celebrating his forty-third birthday), Rosenblatt's Perceptron idea had been sidelined in favor of the more structured DENDRAL-like approaches to artificial intelligence. But in the twenty-first century, the Perceptron is celebrated as a prototype of *machine learning*, which is a catchall term for a vast and expanding library of techniques. None of these require special hardware: in fact, it was always possible to replicate the Perceptron's operation using a general-purpose digital computer. The 1958 demonstration given to the *New York Times* used not a special machine but the US Weather Bureau's computer, in its time off from producing early numerical forecasts. As the power of digital computers rocketed, the need for messy custom hardware with spaghetti wiring evaporated.

Today's machine-learning techniques have mystical-sounding names: it sounds lovely to work with a "support vector machine," take a walk through a "random forest," climb a "gradient-boosted tree," or explore a "convolutional neural network." But these abstract words belie real-world consequences that in some ways surpass Rosenblatt's grandiose visions. Machine learning, through its ability to classify sounds, images, videos, and your internet history or medical records, has enabled a level of industrial, commercial, and state surveillance that is unprecedented in history. The technology is far ahead of attempts to regulate or even understand its consequences.[38] For astronomy, like so many other fields, it has become indispensable; whatever misgiving about its darker side, resisting it is to lock oneself into a twentieth-century enclave.

Astronomy, Science, and Beyond

Ofer Lahav, one of my colleagues at University College London, forged a head start in applying machine learning to astronomy during the late 1990s. After encountering machine learning by chance while on sabbatical in Japan, he worked with a team of students to develop an alternative approach to the problem of measuring redshifts—the crucial third dimension of cosmological maps. Knowing that Bayesian approaches can be fooled when presented with new and unanticipated unknown unknowns, Lahav and his team created a neural network that learned for itself: by showing the network 15,000 galaxies with known, human-confirmed redshifts, the machine was able to predict redshifts for 10,000 more.[39] The technique quickly established a reputation for being fast, practical, and flexible; today, similar approaches are routinely used to generate accurate depth maps for millions of galaxies.[40]

Machine learning's potential extends into every part of astronomy. When the Vera Rubin Observatory scans the sky over the coming decade, it will not just build a static map: it will particularly be searching for objects that move (asteroids and comets) or change brightness (flickering stars, quasars, and supernovae). Cosmologists are particularly interested in supernova explosions because their bright light can be used to trace the way the universe is expanding. Machine learning can be trained to spot these in the ever-changing sky, allowing them to be studied with other, more specialized telescopes before they fade from sight in a matter of a few weeks.[41] Similar techniques can even help sift through the changing brightness of vast numbers of stars to find telltale signs of which host planets, contributing to the search for life in the universe.[42] Beyond astronomy there are no shortage of scientific applications: Google's artificial intelligence subsidiary DeepMind, for example, has built a network that can outperform all known techniques for predicting the shapes of proteins starting from their molecular structure, a crucial and difficult step toward understanding many biological processes.[43]

These examples illustrate why excitement around machine learning has built during this century. There have been strong claims that we are witnessing a scientific revolution. In 2008, Chris Anderson wrote an article for *Wired* magazine that declared the scientific method, in which humans propose and test specific hypotheses, obsolete: "We can stop looking for models. We can analyze the data without hypotheses about what it might show. We can throw the numbers into the biggest computing clusters the world has ever seen and let statistical algorithms find patterns where science cannot."[44]

I think this is taking things too far. Machine learning can simplify and improve certain aspects of traditional scientific approaches, especially where classification (understanding galactic redshifts), processing of complex information (discovering protein shape), or quick action (deciding whether to point telescopes at a potential supernova) is required. But it cannot entirely supplant scientific reasoning, because that is about the search for an improved understanding of the universe around us. Finding new patterns in data is only one narrow aspect of that search. There is a long way to go before machines can do meaningful science without any human oversight.

The Weakness of Data

To understand the importance of context and understanding in science, consider the case of the OPERA experiment, which in 2011 seemingly determined that neutrinos travel faster than the speed of light. The claim is close to a physics blasphemy, because relativity would have to be rewritten; the speed limit is integral to its formulation. Given the enormous weight of experimental evidence that supports relativity, casting doubt on its foundations is not a step to be taken lightly.

Knowing this, theoretical physicists queued up to dismiss the result, suspecting the neutrinos must actually be traveling slower than the measurements indicated.[45] Yet, no problem with the measurement could be found—until, six months later, OPERA announced that a cable had been loose during its experiment, accounting for the discrepancy.[46] Neutrinos traveled no faster than light; the data suggesting otherwise had been wrong.

Surprising data can lead to revelations under the right circumstances. The discovery of Neptune (see chapter 2) is an example. But when the claim is discrepant with existing theories, it is much more likely that there is a fault with the data; this was the gut feeling that physicists trusted when seeing the OPERA results. It is hard to formalize such a reaction into a simple rule for programming into a computer intelligence, because it is midway between the Bayesian and machine-learning worlds. On the one hand, it relates to preexisting knowledge and therefore appears to require a Bayesian approach. On the other, it also relates to unknown unknowns—problems with the experiment that haven't been anticipated in advance—and therefore demands significant flexibility of thought.

The human elements of science will not be replicated by machines unless they can integrate their flexible data processing with a broader corpus of knowledge. There is an explosion of different approaches toward this goal, driven in part by the commercial need for computer intelligences to explain their decisions. In Europe, if a machine makes a decision that impacts you personally—declining your application for a mortgage, maybe, or increasing your insurance premiums, or pulling you aside at an airport—you have a legal right to ask for an explanation.[47] That explanation must necessarily reach outside the narrow world of data in order to connect to a human sense of what is reasonable or unreasonable.

Problematically, it is often not possible to generate a full account of how machine-learning systems reach a particular decision. They use many different pieces of information, combining them in complex ways; the only truly accurate description is to write down the computer code and show how the machine was trained. That is accurate but not very explanatory. At the other extreme, one might point to an obvious factor that dominated a machine's decision: you are a lifelong smoker, perhaps, and other lifelong smokers died young, so you have been declined for life insurance. That is a more useful explanation, but might not be very accurate: other smokers with a different employment history and medical record have been accepted. So what precisely is the difference? Explaining decisions in a fruitful way requires a balance between accuracy and comprehensibility.

In the case of physics, using machines to create digestible, accurate explanations that are anchored in existing laws and frameworks is an approach in its infancy. It starts with the same demands as commercial artificial intelligence: the machine must point to not just its decision (the particular redshift it has guessed for a galaxy, say) but also give a small, digestible amount of information about why it has reached that decision. That way, you can start to understand what it is in the data that has prompted a particular conclusion and see whether it agrees with your existing ideas and theories of cause and effect. This approach has started to bear fruit, producing simple but useful insights into quantum mechanics,[48] string theory,[49] and (from my own collaborations) cosmological structure formation.[50]

These applications are still all framed and interpreted by humans. Could we imagine instead having the computer framing its own scientific hypotheses, balancing new data with the weight of existing theories, and going on to explain its discoveries by writing a scholarly paper without any human assistance? This is not Anderson's vision of the theory-free future of science but a more exciting, more disruptive, and much harder goal: for machines to build and test new theories atop hundreds of years of human insight.

Robot Physicists

The dizzying range of techniques that come under the umbrella of artificial intelligence have one idea in common: attempting to capture some facet of thinking inside a computer program. Before this chapter, I focused on physical simulations—which start with a set of assumptions about galaxies, black holes, or the universe—and turned that into a prediction for observations. By contrast, simulating thought is nearly always about tackling inverse problems: not predicting measurements from theory but inferring backward from measured data to the most likely galaxy redshift, or black-hole collision masses, or density of dark matter in the universe, or even (one day) a new theory entirely.

As it stands, that final goal has not been achieved because only partial aspects of human thinking can be simulated by machines. Despite this, we already see major societal challenges. Artificial intelligence has a profound influence economically (factory workers are being replaced by skilled robots),[51] judicially (police forces are using intelligences that exhibit racial biases),[52] societally (artificial intelligence has enabled new forms of worker surveillance and exploitation),[53] and politically (social-media bots are pumping out propaganda and disinformation).[54]

These examples show how the sci-fi dystopian future where artificial intelligences begin to manipulate and control humans is frighteningly close, if not already here. Computers are taking charge, not in a single Hollywood-friendly spectacular event but through creeping encroachment. There is the potential for deeper destabilization of our world if machines can additionally be coded to be independent, willful, and flexible. Reaching this goal will require profound new breakthroughs, but ultimately it is hard to see why they cannot be achieved. Even if no simplified method will suffice, our own thinking is powered by neurons whose behavior we assume can be described by physics and therefore simulated in a sufficiently powerful computer.

One possible objection is to imagine that quantum effects could play an as-yet unidentified role in our neurons and therefore in thought. But, even then, general-purpose quantum computers are likely to become available sooner or later. Unless brains harness some process outside known physics, I suspect that it is a matter of time before we have the technological capacity to simulate human thought in a comprehensive, convincing way.

As to how much time—years, decades, or more—I wouldn't dare give a firm answer, but, even now, existing systems give a glimpse of what is possible. At the time of writing, one of the most humanlike artificial intelligences available is known uncatchily as Generative Pre-trained Transformer, or GPT, a neural network that has been trained on around 500 billion words taken from the internet, including Wikipedia.[55] A human able to retain all that information would be heralded as a genius.

Yet all that GPT really does with its knowledge is guess at what should come next in a dialogue, like a turbocharged version of the autocomplete on your phone. In its widely-available form, "ChatGPT," it comes across as a bland know-it-all, but the underlying technology can be reconfigured to predict all manner of text. I asked it to mimic the responses of a fictional researcher whom I named The Professor: a renowned expert on quantum gravity who has high hopes for artificial intelligence solving major mysteries in theoretical physics. Here is an extract from the resulting discussion:

Andrew: Why is it so hard to find a theory of quantum gravity?

The Professor: Quantum gravity is a notoriously difficult theory to grasp. There are many theoretical obstacles. In my view, there are also deep conceptual obstacles.

Andrew: How will we overcome those?

The Professor: The theoretical obstacles include intractability— the fact that there are too many possible solutions to the equations, not to mention a great number of other problems such as solving these equations at all.

Andrew: I hope we get past that and actually solve this problem.

The Professor: We cannot hope to solve it directly. However there are some clever ideas that allow us to explore solutions indirectly. Machine-learning techniques can help find solutions that we would not have thought of.

Andrew: So will computers put physicists out of a job?

The Professor: No, computer scientists will put physicists out of a job! In this sense, though, I think we are both working toward the same goals, and perhaps we are working together.

Andrew: Tell me about your dream.

The Professor: My dream is to discover a new theory of quantum gravity that is scientifically convincing and at the same time makes sense to the non-expert.

It's like a mediocre TV script: believable on the surface, but with little substance. GPT's conversations drift aimlessly, its arguments lack any large-scale coherence, and its intelligence is limited to the restatement of existing ideas that it found who-knows-where on the internet. The Professor does not pass my Turing test and, despite claiming to have an admirable dream, is incapable of pursuing even the simplest of goals.

On the other hand, it produces good-quality prose; it can answer basic questions accurately; and it even digs up from somewhere a half-decent in-joke (don't blame computers, blame your computer-scientist colleagues). These kinds of systems, while far from having human-level intelligence, are starting to encroach on the roles of journalists, writers, educators, and even computer programmers themselves—GPT and other similar systems are now able to write code.[56] I have around thirty years until retirement age; it would be unwise to rule out the possibility that, within that time, there could be a sophisticated AI that will code, write, teach, and perhaps even research better than I can.

Out of interest, I asked The Professor whether it was human.

The Professor: Well, at the moment I am a human, but I hope to become a superintelligent computer before long.

I pressed harder, directly accusing it of being an artificial intelligence; it seemed absolutely adamant that I was wrong:

The Professor: You are saying I am an AI? This is the most ridiculous claim I have ever heard.

Andrew: Why is it so ridiculous?

The Professor: You are accusing me of being an AI without providing any arguments in support of this claim. I have been a professor for many years. I have been researching quantum gravity for most of my life. I have been a member of the Stanford faculty for over twenty years. I've published over 200 research papers in physics journals. I don't know what to say.

This is a realistic portrayal of a professor's response to grave challenge: issue an outright, angry denial while seizing the chance to highlight some irrelevant academic credentials. It's irritating, sweet, laughable, and more than a little unnervingly human.

From our position in the real world, you and I can confidently dismiss The Professor's deluded insistence that it has a faculty position at Stanford. We can see what has happened: I asked GPT to play the role of a renowned quantum gravity expert and, during its training, let loose on the internet, it has picked up on (perhaps somewhat justifiable) stereotypes of what such an expert might be like. Combine that with a little creative impetus—which arises naturally from randomness coupled to simple rules applied on vast scales—and you can see how the machine gives the illusion of a personality with jokes, delusions, and all.

It is, at least for now, nothing much more than impressive smoke and mirrors. But our dismissal of The Professor's claims to be human can still come back to bite us: How much more than smoke and mirrors is our own thinking, and are we so sure that we aren't deluded about our own reality? Imagine that a far-future civilization could create, with mind-bogglingly powerful computers, a simulation of the physical world with sufficient detail for artificial intelligence to evolve within it. Going far beyond the limitations of current systems, suppose that such computer-evolved intelligences could match or exceed our own thinking prowess.

It might be hard, but there is nothing in what I have shown you so far that says this would be impossible. Once you buy into the vision, it is only a small jump to start contemplating that we ourselves—you and I—might in fact be intelligences within that simulation, duped into believing that we exist within a physical reality.

Like The Professor, you might take this to be the most ridiculous idea you have ever heard, and I tend to agree. Yet I also think it's at least a little bit spooky and, for that reason, I want to take a close look at the claim.

7

SIMULATIONS, SCIENCE, AND REALITY

In the spring of 1999, aged fifteen, I made a rare trip to the cinema to watch a new film called *The Matrix*, in which a computer programmer discovers his entire life to date has been lived inside a simulated reality. Machines have somehow placed real humans into pods and wired them into a giant game. The remainder of the film traces how the hero, Neo, joins a small group of elite coders who seek to liberate humanity and return everyone to the real world. I distinctly remember this first exposure to the idea that our entire experience is a fraud, because it was deeply unnerving at such a formative age. I left with a sense that reality shouldn't be taken for granted.

Ever since computers reached widespread public attention in the 1950s, science fiction has toyed with the idea that we live in a simulation. Frederik Pohl's short story "The Tunnel Under the World" (1955) provides one prototype in which the consciousness of humans is transplanted into robots who inhabit a purpose-built, tabletop miniaturized city. Stuck within their metaphorically and literally tiny world, the poor souls must repeatedly relive the same day in perpetuity, all so that different product advertisements can be trialed. Each night, a team from the outside world wipes the robots' short-term memory, then resets the environment, providing a precisely controllable testbed for the marketing industry.

Daniel F. Galouye's novel *Simulacron-3* (1964) takes Pohl's idea of a marketing trial and places it fully within a computer, envisioning a company which develops a simulation of an entire city and its population—no need for a miniaturized set. But the scientists who develop the simulation slowly become aware that their own reality is not the true one: it is, itself, a simulation of some "Higher World." This possibility—that all things, including our own bodies and minds, are inside a computer—is known as the *simulation hypothesis*.

It is not just science-fiction writers who find the simulation hypothesis worthy of attention. Computer scientists Edward Fredkin and Konrad Zuse raised it as a serious possibility in the 1950s, and in the early part of this century quantum physicist Seth Lloyd wrote that "a simulation of the universe on a quantum computer is indistinguishable from the universe itself."[1] Public figures like astronomer Neil deGrasse Tyson, physicist Brian Greene, and evolutionary biologist Richard Dawkins all give the simulation hypothesis serious consideration.[2]

Examined closely, there are many different versions of the hypothesis and each of these figures has their own take on it. But a good place to start is with philosopher Nick Bostrom in 2003, who made an argument along the following lines:[3] Assume that, like us, future civilizations are interested in simulating cosmic history, or some fragments of it; one purpose might be to study the formation of the solar system, Earth, and life, or even the evolution and behavior of intelligent organisms. Assume also that computers and simulations continue to increase in their power and refinement. If one grants these two assumptions, humanity (and similarly advanced alien civilizations) will eventually create an ultrasophisticated simulated universe in which intelligent life is implanted or evolves.

Then comes the punch line. Suppose that in the entire past and future of the universe, even a single civilization reaches the required level of technical capability, and that they perform just one simulation. Now you have two possibilities regarding your own existence: either you live inside reality or, potentially, you live inside the simulation. In the latter case, you are a form of artificial intelligence. (This does suppose that artificial intelligences with conscious experiences are feasible, but Bostrom sees no reason to rule it out, and nor do I, nor Seth Lloyd, nor philosopher-of-the-mind David Chalmers.)[4] Given a choice of two explanations—real or simulated—and no way currently to distinguish them, we should seemingly assign a fifty-fifty chance that we are simulated beings.

Bostrom in fact imagines that there may be many advanced civilizations, and that they may want to perform multiple simulations to explore different aspects of history or effects of making changes in the physical laws, much as we do with today's less advanced technology. In that case, the simulated universes hosting life would outnumber the single real one. Suppose ten civilizations each perform ten suitable simulations at any point in their history; your odds of living in the real universe are now a hundred to one against.

It won't have escaped your attention that this is a tower of speculation. Bostrom is careful not to overstate the claim; he agrees that many of the assumptions are contestable. Yet the headline conclusion— the simulation hypothesis itself—has captured the imagination of some great minds to whom the assumptions appear plausible. Society will remain interested in re-creating history: sure, why not? Computers and simulations will continue to increase in power and sophistication: absolutely. Future civilizations will not just settle for a single simulation: no doubt. Consciousness can be understood through science and thereby re-created in a machine: yes, because any other conclusion requires a supernatural view of the mind. To argue against these seems unduly pessimistic, invoking a loss of interest in our origins, or an end to progress in scientific computation, or perhaps an end to civilization itself. Bostrom's point is that, logically, we are faced with a choice. Either we accept that there are stringent limitations on what may be achieved in the future, or we accept the outlandish simulation hypothesis.

Just because a hypothesis seems fantastical doesn't mean that we should reject it out of hand. Physics is full of absurdities: time that runs at different rates, particles that are in many places at once, expanding universes, and the like. One should try to keep an open mind. Yet the simulation hypothesis isn't just absurd. It's explosive. It is a type of religion, seemingly inferred from science; it states that there is an architect of our universe, one who might have agency to intervene in the course of history. Despite this, Richard Dawkins, an avowed atheist, admits that Bostrom makes a plausible argument.[5] (According to Dawkins, the creators of the simulation themselves arose through evolutionary processes and therefore shouldn't be classed as gods; that opens up questions about the definition of a god but, regardless of their origins or names, the creators' power over our reality would be prodigious.)

By intertwining religion with science and technology, the simulation hypothesis has a volatility which has ignited plenty of interesting discussion, although I think most of the discourse skates over detail too lightly. Hidden in Bostrom's postulates are a lot of presumptions concerning what scientists aim to achieve by building simulations. Even if science and computation continue to progress, and even if humanity retains its curiosity about our origins and behavior, simulations that replicate reality in detail are not necessarily the result. Conversely, if incredibly detailed simulations *are* attempted, they would be so different from the kind of simulations that we perform today, and they would be hosted by civilizations with such different capabilities and intentions, that we should not presume our reasoning about them carries a clear meaning. To flesh out these ideas, I am going to revisit lessons from previous chapters about why simulations are useful, and how they are performed in practice.

But if the relevance of simulations as a literal explanation for our reality may have been exaggerated, their revolutionary implications for science have certainly been underplayed. Science is the journey toward a better understanding of nature, and simulations are a new phase. The scientific process has been refined over several centuries, but simulations have only had a few decades to find their feet; there is more to be understood about the various roles they can play. Sometimes they look like theoretical calculations, sometimes like empirical experiments, and sometimes like an entirely new way of building a collaborative, human vision of the universe.

By understanding where the simulation hypothesis is weakest, one begins to appreciate where simulations themselves are strongest and most radical, and where they might be headed in the future. In this final chapter, I want to explore these tensions and reach a better understanding of what simulations are really about.

The Resolution of Reality

The simulation hypothesis presupposes vast future improvements in the quality of our digital realities, and one metric on which this can be assessed is resolution. In the context of weather, resolution loosely corresponds to the number of grid squares, the more the better; for dark matter or galaxy simulations, a good starting point is the number of smarticles and, again, the more the better. State-of-the-art cosmological simulations contain around 20 billion smarticles and associated with each are at least six numbers (three for position, three for motion, and sometimes others to represent factors such as chemical composition). Each number can in turn be decomposed into bits, the basic unit of computer storage. Adding it all up, the number of bits associated with the largest cosmological simulations performed to date is in the order of 100 trillion, or 10^{14}. By contrast, I estimate the number of bits in the Richardsons' weather simulation at 1,000, or in Holmberg's light-bulb simulation at 3,000.[6] Humanity has certainly made enormous progress, but this needs to be put into the context of a mind-bogglingly detailed universe.

One can estimate an equivalent number for the resolution of reality itself, although the way to do so cannot be based purely on particle number. That is because quantum mechanics, with its fuzzy uncertainties, permits particles to pop in and out of existence continually, and for them to develop entanglement, the interdependencies discussed in chapter 5. All this bumps up the storage requirements in a way that cannot easily be calculated directly, necessitating a different approach, one that will count the number of *qubits*—the units of storage in a quantum, rather than a classical, computer.

The qubit content of reality is calculated in a rather surprising way. First, we estimate the total energy contained within the observable universe, which can be extrapolated based on the cosmic microwave background and other observations. (According to relativity, mass is just another form of energy, so all the matter in the universe must be included in this figure, too.) Next, we work out how many qubits are required to represent all possible ways that such a vast amount of energy could behave over the course of cosmic history. This sounds like a vague and therefore impossible task, but in the late 1970s, Stephen Hawking and Jacob Bekenstein derived formulae that allow it to be achieved. The key step is to imagine the entire observable universe being swallowed by a giant black hole; if this happened, all the qubits in the universe would be lost. As I previously mentioned, black holes appear to destroy information about the particles that enter them; while there is controversy among physicists about whether the information is truly lost forever, this doesn't affect the qubit-counting argument.[7] Adding up the lost qubits gives an estimate of how many were in the universe in the first place, which turns out to be 10^{124} qubits.[8]

Evidently, this is stupendously large compared to 10^{14} classical bits in our simulations. Not only that, but a classical bit is far less powerful than a quantum one. Today's quantum computers are impressive machines, but they do not have the precision needed to realize Feynman's dream of a universal simulator, even for a small number of qubits.[9] There could hardly be a greater mismatch between today's simulations and the steep requirements for creating a perfect mimic of reality.

A *perfect simulation hypothesis*, in which we are inside a faultless or near-faultless re-creation of a parent reality, is therefore untenable. Even if one accepts an extreme extrapolation to the capabilities of future hardware, storing 10^{124} qubits will still be impossible. This follows by reversing the flow of the previous calculation; instead of asking how many qubits are needed to represent a certain amount of energy, one asks how much energy a quantum computer would require to represent a desired number of qubits. The overall calculation is then circular: it goes from energy to qubits and back to energy again, implying that one would need to use all the energy in the real universe in order to create a single simulated universe. Even if it were possible, this is obviously pointless, not to mention unethical.

There is therefore no hope of ever performing whole-universe simulations at the resolution of reality itself. So what of Seth Lloyd's contention that a quantum simulated universe is indistinguishable from reality? This is still fine, provided it is seen as a statement of principle rather than one of practice. Such thinking doesn't imply the Bostrom-style simulation hypothesis and in fact has its own name: the *it-from-qubit hypothesis*. Reality, "it," is equivalent to the processing of quantum bits of information, "qubits."[10]

The it-from-qubit hypothesis and the simulation hypothesis, while related, are strikingly different. It-from-qubit is an observation that our universe seems analogous to a giant quantum computer; one can choose to think of reality as a "simulation," but the word is being used metaphorically because there is no suggestion of any hardware on which this simulation is being performed. The purpose of setting up such an analogy is to provide a way to think about physics, in the hope that it will lead to tangible progress in difficult areas such as quantum gravity. The it-from-qubit hypothesis is therefore epistemological: it provides a way to think about the nature of our scientific theories.

The simulation hypothesis, on the other hand, proclaims itself as ontological: a way to think about the nature of reality itself, as being contingent on a machine and creator in a universe that transcends our own. This is an extra layer on top of it-from-qubit; it says that reality is not only analogous to simulations but that it is literally a simulation. If this were true, the reality in which the simulation is being performed must contain many more qubits than our own to enable the computer and its creators to fit. I would not remotely trust our ability to reason sensibly what creatures in such a universe might be up to, much less to count how many simulations they might choose to perform.

Does Reality Have a Sub-grid?

So far, I have ruled out the possibility that a perfect simulation of the universe could be performed within the universe. But what about an imperfect simulation, in which only part of the universe is reproduced with high fidelity, and the remainder is somehow approximated? This would cut the computational cost; let's call it the *budget simulation hypothesis*.

At first sight, budget simulations are a natural extrapolation from the kind of simulations we perform today. Simplifying physics is an integral part of our process, so central that we even have a name for it: sub-grid. Whenever detail is important but cannot be captured for lack of resolution—the formation of clouds and rain in weather, or the behavior of stars and black holes in galaxies, for example—simulation designers add sub-grid rules that crudely mimic what is missing. On top of this, cosmologists don't necessarily use a fixed resolution across our simulations; when we are interested in understanding how galaxies form, we often select just one or two for the computer to track in maximal detail, while the remainder of the virtual universe is assigned minimal attention. For some far-future simulation, this might translate into a single solar system where physics is calculated with high precision, while farther-flung parts of the universe are re-created only in outline, making heavy use of sub-grid prescriptions.

Let's for a moment assume we are inside such a simulation. There are layers to the simulated reality: an inner core where we reside, and an outer part which is vastly simplified into a set of sub-grid rules, acting almost like CGI scenery in a movie. But the simulation must have been engineered to prevent us from noticing the boundary between full and simplified physics with humanity's collected experiments and observatories, not to mention our gravitational wave, neutrino, or cosmic-ray detectors—as far as we can see, physics out in space seems exceptionally like physics here on Earth. Whenever a leap in our technology takes place, like the first detection of gravitational waves or the construction of a more powerful telescope, there is the possibility that we reach the level of sophistication required to notice the difference between inner reality and outer CGI.

So far we have never detected such a mismatch, implying the putative simulators must have anticipated our experiments and measurements and crafted sub-grid prescriptions which are resistant to this line of attack. To me, this seems like a conspiracy theory. But some physicists, including the outstanding cosmologist John Barrow, have taken the opposite view, claiming that what I am calling the budget simulation hypothesis is a scientific, testable proposition. The test for living in such a simulation would be to search for the inaccuracies, something like "glitches in the Matrix."[11]

I am doubtful of this claim. Even if experiments or observations did find what looks like an inaccuracy, nobody could be sure that it wasn't just some new, still-to-be-understood but natural phenomenon. Discoveries along such lines might be exciting, or they might be a repeat of the OPERA faster-than-light debacle, but in neither case would they demonstrate that we live in a simulation. At the heart of the problem is that unless we are confident about the *purpose* of a budget simulation, we cannot meaningfully speculate about which sub-grid simplifications would be acceptable to its creators. I fear otherwise that looking for glitches is an exercise in data analysis divorced from any clear theoretical underpinning, and I argued in chapter 6 that such exercises are not science.

Contrast glitch-searching with the admittedly speculative ideas in mainstream cosmology, like inflation, dark matter, and dark energy. Each of these three pillars has clear motivations and leads to expectations that can be tested, as I discussed in earlier parts of the book. There are malleable and dubious aspects of them, and they may one day be supplanted with refined ideas; but they are working theories with a core premise that can be used to set expectations and motivate specific observations or experiments.

Conversely, when we examined close-up, the core premise of the budget simulation hypothesis remains undefined. If the hypothesis is to be anything other than a conspiracy theory, someone will need to clearly elucidate the proposed future scientists' purpose. That is the starting point for understanding the kind of sub-grid that would be in play, and therefore the kind of tests we might construct to reveal them.

Personally, I doubt that future civilizations will bother themselves with building simulations compatible with the simulation hypothesis in any form. It is not because I am pessimistic enough to feel we will lose our curiosity or research capacity in the future. It is rather because I am optimistic that we will direct our curiosity and technological prowess toward something more interesting. To understand more, it is time to revisit what our contemporary simulations have achieved.

Simulations as Calculations

Scientists of the future might not attempt ultradetailed re-creations of the universe, but they may very well lean heavily on simulations of various kinds. Science has always progressed in lockstep with technology, but the values at its core have changed little since the Enlightenment. Seventeenth-century philosopher Francis Bacon, concerned that our sensory experiences are subjective and might therefore lead to starkly wrong conclusions, suggested using careful experimentation to correct misapprehensions over time. By performing as many controlled experiments as possible, by generalizing the results to draw tentative conclusions that may be revised later, and by sharing the resulting knowledge widely, he believed that humanity could increase its understanding of, and power over, nature.

History proved him correct. As a cosmologist, I am bound to add that experiments are not always possible; being unable to control the universe at large, we sometimes need to lie back and observe what nature chooses to reveal through light arriving from the farthest reaches of space. Nonetheless, the essence of hypothesis-driven science survives because we can make predictions for what we will see in a particular new telescope and test those. Where do simulations belong within this process of hypothesizing and testing?

Consider the Large Hadron Collider (LHC), the giant international physics experiment in Geneva; its approach is to smash particles together and see what emerges. Two snowballs colliding at low speed will probably just stick together or bounce, but at high speed they shatter into powder. Something similar happens in subatomic physics, although one has to add a layer of quantum weirdness: due to particles being bundles of energy inside quantum fields, the fragments that emerge from the collision can be different from the particles that collided. It is as though these particular snowballs can unpredictably turn into sugar, flour, or powder paint at the moment of impact.

The LHC is most famous for finding Higgs bosons in the debris of proton collisions. Simulations were central to that search. There is no such thing as a detector for the Higgs because it is a momentary, unstable by-product of the collision and rapidly disintegrates into exotically named particles like quarks and gluons. Confusing matters further, all of these are produced in the original collision itself, regardless of whether a Higgs boson is involved. Working out what really happened is an inverse problem, forcing physicists to compare experimental data with different scenarios—what would have been seen with and without the influence of a Higgs boson. Calculating for each case how many particles of what type and energies will emerge and be detected is impracticable for a human; computer simulations are used to produce the predictions and are therefore implicated in the discovery itself.

Earlier chapters covered similar examples. When gravitational waves are detected, their particular shape must be compared with simulations of colliding black holes or neutron stars to interpret what actually happened. Similarly, comparing observations of cosmic large-scale structure seen through telescopes with predictions from simulations showed that dark matter cannot be composed of neutrinos.

From this perspective, simulations are a conduit within the process of science. They do not provide a hypothesis; this is given by the underlying theory. They do not provide data; this is given by the specific experiment or observation. Rather, they provide the connection between these two, by predicting what the data *would* have been under each hypothesis. There is a complication in that the prediction is always approximate; as I recapped above, we are forced to simplify to make the computation tractable. Understanding whether the approximations are distorting a particular comparison is the art of simulation.

Scientists throughout history have performed simplifications that are not fully justified in order to make contact with reality. Consider Einstein's theory of general relativity: it is one thing to write down the abstract equations that describe how space-time is warped in response to material, but quite another to decide what practical implications this has. Einstein calculated that the orbit of Mercury would differ slightly from that implied by Newton's theory, and the updated prediction matched the known trajectory of the planet almost perfectly.[12] He also predicted how light would be deflected as it passed massive objects—the effect known as *lensing*—and this, too, was later confirmed.[13] In formulating these predictions, Einstein made use of an approximation scheme which was somewhat questionable, and only later justified once Schwarzschild completed his more accurate calculation of gravity around stars.

So calculating the consequences of a theory—even approximately—has always been an essential part of science. The distinction between simulations and manual calculations seems marginal when framed in this way. Richardson's manual weather forecast and Holmberg's lightbulb galaxies demonstrate that there is nothing a computer can do that a human team couldn't accomplish, given inexhaustible time and patience.

Simulations as Experiments

Is that all there is to it? I don't think so, and neither do some philosophers of science (notably Margaret Morrison) who suggest that computer simulations have much in common with experiments.[14] At first
glance this claim seems questionable: the entire point of a simulation
is that it does precisely what the programmer asks of it; conversely,
the point of an experiment is that it does what nature demands of it.
Even if we believe that our theories might describe nature's laws ad-
equately, simulations involve substantial approximations to those
laws, whereas experiments take place within the real universe.

Yet experiments are also not entirely down to nature. They are
devised and orchestrated by humans, using processes that often involve approximations and assumptions laden with our own theories
and expectations. Suppose you are designing a new aircraft wing and
wish to understand how it will behave in a certain airflow. To work
it out, you can either make a scale model and place it in a real wind
tunnel or construct a digital simulation. In the former case, you are
making the approximation that airflow should behave similarly
around a reduced-size wing as around a full-scale equivalent and, additionally, that the proximity to the edges of a wind tunnel will have
little bearing on the overall results. In the simulation case, you are
assuming that air behaves as a fluid, and that various approximations
in the computer code will not disrupt your conclusions. Both are powerful but flawed tests, and it is not clear why you would label one of
them an experiment but not the other.

Experiments are supposed to teach us something that we didn't
know before. By any normal standards we can definitely learn new
facts about physics using simulations: whether a given wing design

will work well is, after all, a fact. Yet one could regard the performance of a particular aircraft wing as a detail; if physicists are confident that we already possess the theoretical knowledge to describe airflow, all a simulation does is churn that knowledge into a new form. In this view, simulations are helpful tools to reveal hidden truths, but will not teach humanity anything fundamentally unknown. Can a simulation teach us anything truly new which is not already implied by our existing knowledge?

The answer to this question depends on your attitude toward physics. One way to envision physics is as a search for a final *Theory of Everything*: a single, coherent framework that describes all the different particles and forces in our universe. At present, this remains elusive, in large part because gravity behaves so differently from all other forces. If a Theory of Everything is the goal, simulations' role can only be to trace the consequences of a given proposal through to its implications for experimental data, in the way I outlined for the Higgs boson, black holes, or dark matter.

But physics is not just about finding a Theory of Everything. Many phenomena that we care about—in the everyday world and in the cosmos beyond—are almost independent of the underlying laws of physics. Take *thermodynamics*, which is the theory that allows us to bundle countless atoms and molecules into a single parcel of air. It is immensely powerful, describing why cups of hot tea cool, how to build engines for optimal efficiency, and why life in the universe cannot last forever.

Thermodynamics contains concepts like heat and entropy. These have no meaning within particle physics; they come into play only when we consider large numbers of particles and impose additional layers of interpretation on how these particles are behaving collectively. While the particular way in which atoms or molecules behave does have a bearing on these large-scale properties, the relationships are loose.

When teaching thermodynamics to undergraduates, I provide them with a code which simulates the way that particles zip around. The simulation doesn't incorporate details of these particles in a careful way—compared with reality, they are far too heavy, not numerous enough, and unable to vibrate or rotate in the way that actual molecules do. And yet, the basic laws of thermodynamics can still be discovered within these simulations. The students set up scenarios with hot and cold regions (meaning the particles are moving at respectively higher and lower speeds) and find that the temperatures equalize. They encase particles in a virtual box that changes its shape over time, discovering how gases heat up when compressed and cool down when released, the same principle upon which a refrigerator is based. They study how particles of gas diffuse, quickly coming to permeate an entire room even if they are initially concentrated in one corner. All these insights can be obtained from a simple simulation, despite the fact that the details of the particles' behavior are wrong.

Phenomena that emerge in this way are absolutely discoverable within simulations, just as they are in a traditional experiment. Thermodynamics is a slightly artificial case, because it was understood well before the invention of the computer. But I have given you plenty of genuine examples: the way that black holes suddenly turn on their galaxies, the way that exploding stars within galaxies can slowly resculpt the dark matter around them, or how random networks of virtual neurons can learn in a way reminiscent of biological brains.

The underlying physics of black holes, dark-matter particles, super-nova explosions, or brain cells need not be perfectly reproduced because it is the emergent behavior that matters.

The detail that we can afford to include in simulations is pitiful compared with that offered by reality. Nobody forms planets inside their computer galaxies; it is a rare thing to be able to pick out individual stars; the rich phenomenology of black holes is reduced to a short list of rules about how they swallow gas and create energy. As a result, the minutiae of any new behaviors will surely not be quite right, but we still capture something of their essence.

Simulations are therefore a laboratory in which scientists can experiment and learn. Sometimes the computational necessity to replace real physics with a simplified caricature can actually make simulation-based experiments more powerful, because the goal is not to reproduce but to understand how nature has sculpted our cosmos. A good way to accomplish that is to tinker with bits of physics: if you want to understand what effect black holes are having on your galaxy, try suddenly throwing a switch which prevents them from devouring any more gas. My collaborators and I recently did exactly this, finding that galaxies which had long since stopped forming any new stars suddenly sprang back to new life when the destructive effects of the black holes were discontinued.[15]

This experiment was only possible because the black-hole effects were neatly contained in the sub-grid rules, themselves expressed in a single computer file within the simulation code. Consider a hypothetical future simulation in which resolution has increased to the point that sub-grid rules about black holes are unnecessary. On the face of it, this would be a positive development, bringing the simulation closer to reality. Instead of best guesses about how quickly a black hole sucks material from its surroundings, and how much energy it deposits, the relevant underlying physics—general relativity, particle physics, magnetic fields, and so on—would be fully in charge. But if that's the case, it will no longer be possible to neatly isolate the effects of black holes alone, because one couldn't disable any of these fundamental components without having profound side effects.

So there is a fine balance between the strengths and drawbacks of approximate, relatively simple sub-grid-based simulations. Adding ever more detail may be unattractive to future generations of scientists if they value something which has comprehensible outcomes. If simulations are experiments, they are created with the goal of learning about, rather than re-creating, the universe.

Simulations as Science

Simulations are calculations, allowing us to trace the consequences of physics for Earth's atmosphere, or a galaxy, or the whole cosmos. They are experiments, showing us how complex behavior emerges from simple rules. They are tools that have shaped our modern life, through developments like numerical weather forecasting and artificial intelligence. However, they are not facsimiles of our reality, and it is unlikely that they ever will be.

The popularity of simulations among cosmologists—a dozen or so separate codes, hundreds of simulations and scientific papers each year, and scores of eye-catching press releases—doesn't mean that we are working toward a final, perfect simulation of everything. As I have argued in this chapter, such an accomplishment of extreme detail is impossible and quite probably pointless.

Rather, simulations offer a way to structure scientific knowledge, insight, and collaboration. No individual could construct a simulation and compare it to data from a modern survey telescope. The range of expertise is too vast, covering hydrodynamics, star formation, black-hole formation and growth, quantum mechanics, optics, and artificial intelligence. One can study any of these for a lifetime and still have more to learn.

This open-endedness is what makes physics exciting, and working with simulations even more so. But it also means that collaboration is essential. I didn't appreciate this point when I was young: computers offered an escape from the human world where I felt permanently awkward. They were a portal into another reality where pure thought comes to life. Nobody bothered me too much if I chose to inhabit this world; my school-leaving yearbook mentions my main achievement as "managing to construct and live within his own universe."

I cannot now remember whether I expected professional simulations to be a scaled-up version of this computer nerdery, where a lone individual can lock themselves away and create their own world. If that was my expectation, it was gloriously wrong: it is the human factor that makes simulations what they are. Since the Enlightenment, working together has been at the heart of science, because collaboration can far extend the abilities of a single human mind. Over time, a flawed but effective system evolved in which scientific ideas are disseminated in scholarly journals. Through these journals, and especially with digitized online archives, a library can offer access to almost the totality of human knowledge.

Being able to access the publications is one thing, but digesting and understanding the knowledge is entirely another. Well-written computer code changes the demands of the scientific process; rather than a single individual needing to absorb all that information, different specialists work in tandem, distilling their knowledge into pieces of code that combine within an overarching structure. That this is even possible vindicates Grace Hopper's push toward human-readable coding. Different aspects of the simulation are described in their respective files, and the machine takes care of combining them into a single, long list of instructions in its own ultradetailed language. That way, the parts of the code that govern, for example, formation of new stars, or the behavior of black holes, or the generation of quantum initial conditions can be changed or replaced without touching other sections.

The most exciting results from simulations are not the virtual worlds they generate, which are only ever a poor shadow of reality. Simulated worlds in themselves are no more thrilling than a weather forecast. The exhilaration lies in the human domain, where simulations express and explore relationships between different scientific ideas. Code is a set of instructions to the computer, but it is also a living, evolving, collective expression of how we think about the universe, combining different people's ideas on one canvas.

The most enjoyable part of my work is collaborating with other humans to understand the results that computers produce—we visualize, interrogate, interpret in an effort to transform simulated worlds into knowledge about our own reality. The stories that cosmologists have extracted from simulations are already part of cosmological orthodoxy, explaining how our planet's formation began when galaxies condensed out of a gargantuan cosmic web of dark matter, itself sculpted by gravity from nothing more than microscopic quantum mechanical ripples.

These insights are major contributions to cosmology by any measure. The accomplishment isn't literally to re-create the universe but to understand how complex phenomena emerge even when the rules coded into the simulation are individually simple. Studying this kind of emergence was all but impossible a few decades ago; by the centuries-long standards of scientific progress, the whole idea of simulations is an infant. Give it another few decades, and let's see what more there is to discover.

ACKNOWLEDGMENTS

I am grateful to the many brilliant scientists and research students I have worked with, all of whom have influenced my thinking. I am especially indebted to my longest-term collaborators Fabio Governato, Hiranya Peiris, and Justin Read, alongside my PhD supervisors Max Pettini and Anthony Challinor. Hiranya has been a steadfast and inspiring colleague through some difficult years and was also kind enough to read a very early draft of this book. While writing and editing I have received valuable help from Jonathan Davies, Ray Dolan, Richard Ellis, Carlos Frenk, Gandhali Joshi, Matthew van der Merwe, Joe Monaghan, Claudia Muni, Ofer Lahav, Luisa Lucie-Smith, Michael May, Julio Navarro, Tiziana di Matteo, and Simon White, as well as the Smithsonian Libraries and Archives and the Niels Bohr Library and Archives. Much of my own research that I describe was supported by grants from the European Research Council, the Royal Society, and the Science and Technology Facilities Council.

My editors David Milner, Michal Shavit, and Courtney Young have all been immensely supportive, patient, and insightful. Chris Wellbelove, my agent, was instrumental in conceiving and shaping the book; without his encouragement and help in drawing together various threads, I would never have started to write. The title was suggested by Jamie Coleman. My approach to talking about science has been sculpted by the presenters, producers, and directors that I've been lucky enough to work alongside on various projects, particularly Helen Arney, Matt Baker, Jonny Berliner, Hannah Fry, Timandra Harkness, Delyth Jones, Michelle Martin, Jonathan and Elin Sanderson, Alom Shaha, Tim Usborne, and Tom and Jen Whyntie.

This book is for my family. My parents, Libby and Peter, have always encouraged and supported me, and I couldn't have asked for a better big sister, Rosie. Most of all, the love and kindness of my wife, Anna, and son, Alex, make living in this universe worthwhile. I am sorry about all the weekends.

NOTES

INTRODUCTION

1. SatOrb was originally published in a book titled *ZX Spectrum Astronomy: Discover the Heavens on Your Computer* by Maurice Gavin (1984), Sunshine Books. I assume my father, or one of his friends, had typed out the program and then saved it to tape, a common pastime in the early days of home computing.
2. Garnier et al., *PLOS Computational Biology* 9, no. 3 (2013): e1002984.
3. Deneubourg et al., *Journal of Insect Behavior* 2, no. 5 (1989): 719.
4. The 2021 capacity was around 8,000 exabytes, or 6×10^{22} bits. The total mass of the atmosphere is around 5×10^{18} kg, which translates to around 10^{44} molecules. So you would need 10^{21} times more storage for one bit per molecule. Redgate & IDC (September 8, 2021), in *Statista*, from statista.com/statistics/1185900/worldwide-datasphere-storage-capacity-installed-base, accessed July 10, 2022.
5. "They Tried to Outsmart Wall Street," *New York Times*, March 10, 2009.
6. Tankov, *Financial Modelling with Jump Processes* (Chapman & Hall, 2003).
7. Mandlebrot, *Journal of Political Economy* 5 (1963): 421.
8. Derman, *Models. Behaving. Badly.: Why Confusing Illusion with Reality Can Lead to Disaster, on Wall Street and in Life* (Wiley, 2011).

CHAPTER 1: WEATHER AND CLIMATE

1. Moore (2015), *The Weather Experiment*, Chatto & Windus; Hansard HC Deb., June 30, 1854, col. 1006.
2. Gray (2015), *Public Weather Service Value for Money Review*, Met Office; Lazo et al., *Bulletin of the American Meteorological Society* 90, no. 6 (2009): 785.

3. Pausata et al., *Earth and Planetary Science Letters* 434 (2016): 298.

4. Wright, *Frontiers in Earth Science* 5 (2017), doi.org/10:3389/feart.2017:00004.

5. "The Smithson Institute,"*New York Daily Times*, November 2, 1852.

6. *Annual Report of the Board of Regents of the Smithsonian Institution* 32 (1858).

7. *Chicago Press & Tribune*, August 15, 1959.

8. *New York Daily Times*, December 14, 1854.

9. Landsberg, *The Scientific Monthly* 79 (1954): 347.

10. *The Times* (London), "The Late Gale in the Black Sea," January 26, 1863.

11. *The Times* (London), "Admiral Fitzroy and the Weather," April 11, 1862.

12. Humphreys, US National Academy of Sciences, *Biographical Memoirs* 8 (1919): 469.

13. Humphreys, *Biographical Memoirs*.

14. Abbe, *Monthly Weather Review* 29, no. 12 (1901): 551.

15. Stevenson, *Nature* 400 (1999): 32.

16. Technically, the list of numbers might not need to be literally endless, because there are limits to the minutiae. Once every fleck of foam in the crashing wave has been captured, it might be possible to stop adding detail. But this is still far out of reach.

17. Humphreys, *Biographical Memoirs*.

18. In the preface to his 1922 book, Richardson mentions that "the arithmetical reduction of the . . . observations was done with much help from my wife," Dorothy, prior to his posting in France. Since this is a crucial and challenging part of the work, by today's standards Dorothy would be counted a coauthor, even if Lewis Fry wrote the words. Lewis Fry Richardson (1922), *Weather Prediction by Numerical Process*, Cambridge University Press, reprinted 2006, with an introduction by Peter Lynch.

19. Lynch, *Meteorological Magazine* 122 (1993): 69.

20. Richardson, *Weather Prediction* 219; Ashford, *Prophet—or Professor? The Life and Work of Lewis Fry Richardson*, (Adam Hilger, 1985).

21. Lynch, *Meteorological Magazine*.

22. Siberia, 1968, according to *Guinness World Records*, guinnessworldrecords.com/world-records/highest-barometric-pressure, accessed October 28, 2022.

23. Lynch, *The Emergence of Numerical Weather Prediction* (Cambridge University Press, 2014).

24. Richardson, *Weather Prediction*, 219.

25. Durand-Richard, *Nuncius* 25 (2010): 101.

26. "Pascaline," *Britannica Academic*, October 7, 2008, academic.eb.com/levels/collegiate/article/Pascaline/443539, accessed July 20, 2022.

27. Freeth, *Scientific American* 301(2009): 76.

28. Babbage, *Passages from the Life of a Philosopher* (Longman, 1864), 70.

29. Fuegi and Francis, "Lovelace & Babbage and the Creation of the 1843 'Notes,'" *IEEE Annals of the History of Computing* 25, no. 4 (2003): 16.

30. Fuegi and Francis, "Lovelace & Babbage."

31. Friedman, "From Babbage to Babel and Beyond: A Brief History of Programming Languages," *Computer Languages* 17, no. 1 (1992): 1.

32. Fuegi and Francis, *Computer Languages*.

33. Lovelace, reprinted in *Charles Babbage and His Calculating Engines: Selected Writings by Charles Babbage and Others* (Dover Publications, 1963), 251.

34. Fuegi and Francis *Computer Languages*.

35. "ENIAC," *Britannica Academic*, 31, January 2022, academic.eb.com/levels/collegiate /article/ENIAC/443545, accessed October 29, 2022.

36. John von Neumann, in *Fortune* magazine (1955), reprinted in *Population and Development Review* 12 (1986): 117.

37. Haseltine, "Cold War May Spawn Weather-Control Race," December 23, 1957, *Washington Post* and *Times Herald*.

38. Harper, *Endeavor* 32, no. 1 (2008): 20.

39. Fleming, *The Wilson Quarterly* 31, no. 2 (2007): 46.

40. Fleming, *The Wilson Quarterly*.

41. Charney, Fjörtoft and von Neumann (1950), *Tellus*, 237.

42. Williams, *Naval College Review* 52, no. 3 (1999): 90.

43. Hopper (1978), *History of Programming Languages*, Association for Computing Machinery, 7.

44. Hopper (1978), *History of Programming*.

45. Hopper (1978), *History of Programming*.

46. Quoted in Platzman, *Bulletin of the American Meteorological Society* 49, (1968): 496.

47. Bauer, Thrope and Brunet *Nature* 525 (2015): 47.

48. Alley, Emanuel & Zhang, *Science* 363, no. 6425 (2019): 342.

49. McAdie, *Geographical Review* 13, no. 2 (1923): 324.

50. Smagorinsky and Collins, *Monthly Weather Review* 83, no. 3 (1955): 53.

51. Coiffier, *Fundamentals of Numerical Weather Prediction* (Cambridge University Press, 2012).

52. Lee and Hong, *Bulletin of the American Meteorological Society* 86, no. 11 (2005): 1615.

53. Princeton University press conference, October 5, 2021, youtube.com/watch?v =BUtzK41Qpsw, accessed 28 October 2022.

54. Judt, *Journal of the Atmospheric Sciences* 77 (2020): 257.

55. Hasselman, *Tellus* 28, no. 6 (1976): 473.

56. Jackson, *Notes and Records* 75 (2020): 105.

57. John von Neumann (1955), in *Fortune* magazine, reprinted in *Population and Development Review* 12 (1986).

58. Morrison, *Scientific American* 226 (1972): 134.

59. IPCC, *Climate Change 2021: The Physical Science Basis*. Contribution of Working Group I to the Sixth Assessment Report of the Intergovernmental Panel on Climate Change (Cambridge University Press, 2021).

60. Manabe and Broccoli, *Beyond Global Warming* (Princeton University Press, 2020).

61. The Intergovernmental Panel on Climate Change ranks this as a hugely important cross-check. IPCC *Climate Change, 2021: The Physical Science Basis*, Sec. 3.8.2.1.
62. *Daily Mail*, "Why Weren't We Warned?" October 19, 1987.
63. *Daily Mail*, "Fish Gets Icy Blast from Neighbours," October 19, 1987.
64. Alley et al., *Science* 363, no. 6425 (2019): 342.
65. These are the relative contributions by volume. When calculated by mass (which allows a more direct comparison with figures for the universe) the numbers are instead 75 percent and 23 percent respectively. Walker, *Evolution of the Atmosphere* (Macmillan, 1977).

CHAPTER 2: DARK MATTER, DARK ENERGY, AND THE COSMIC WEB

1. Hays, Imbrie, and Shackleton, *Science* 194 (1976): 4270.
2. James Lequeux, *Le Verrier—Magnificent and Detestable Astronomer*, translated by Bernard Sheehan (Springer, 2013).
3. Davis, *Annals of Science* 41, no. 4 (1984): 359.
4. Davis, *Annals of Science*.
5. Rudolf Peierls, *Biographical Memoirs of Fellows of the Royal Society* 5174 (1960).
6. Peierls *Biographical Memoirs*.
7. Pauli (1930), letter to Gauverein meeting in Tübingen, web.archive.org/web /20150709024458/https://www.library.ethz.ch/exhibit/pauli/neutrino_e.html.
8. Lightman and Brawer, *The Lives and Worlds of Modern Cosmologists* (Harvard University Press, 1992).
9. Rubin, *Annual Review of Astronomy and Astrophysics* 49 (2011): 1.
10. Rubin, *Annual Review of Astronomy*.
11. Quoted in Bertone and Hooper, *Reviews of Modern Physics* 90, 045002 (2018).
12. Lundmark, *Meddelanden fran Lunds Astronomiska Observatorium*, Series I, 125 (1930): 1.
13. F. Zwicky, *Helvetica Physica Acta* 6 (1933): 110; Zwicky, *Astrophysical Journal* 86 (1937): 217.
14. Rubin, *Annual Review of Astronomy*.
15. Holmberg, *Astrophysical Journal* 94 (1941): 385.
16. Holmberg, *Meddelanden fran Lunds Astronomiska Observatorium*, Series II, 117 (1946): 3.
17. Lange, *Naturwissenschaften* 19 (1931): 103–7.
18. Rood, *Publications of the Astronomical Society of the Pacific* 99 (1987): 943.
19. White, *Monthly Notices of the Royal Astronomical Society* 177 (1976): 717; Toomre and Toomre, *Astrophysical Journal* 187 (1972): 623–66.
20. White, *Monthly Notices*.
21. Geller and Huchra, *Science* 4932 (1989): 897–903.

22. Interview of Marc Davis by Alan Lightman on October 14, 1988, Niels Bohr Library & Archives, American Institute of Physics, aip.org/history-programs/niels-bohr -library/oral-histories/34298.

23. "Cosmic Extinction—The Far Future of the Universe," Durham University Global Lecture Series, June 7, 2022.

24. Frenk, private communication (2022).

25. Actually, some early experimental results erroneously pointed to a greater neutrino mass than we now know to be allowable. Even those incorrectly large masses were tiny compared to atoms. Lubimov, et al., *Physics Letters B* 94 (1980): 266.

26. Peebles, *Astrophysical Journal* 258 (1982): 415.

27. White, Frenk, and Davis, "Clustering in a Neutrino-Dominated Universe," *Astrophysical Journal* 274 (1983): L1.

28. Aker et al., *Physical Review Letters* 123, 221802 (2019).

29. Silk, Szalay and Zel'dovich, "The Large-Scale Structure of the Universe," *Scientific American* 249, no. 4 (1983): 72.

30. White, private communication (2021).

31. Interview of Marc Davis by Alan Lightman (1988).

32. Lightman and Brawer, *Lives and Worlds of Modern Cosmologists*.

33. Huchra, Geller, de Lapparent, and Burg, *International Astronomical Union Symposium Series* 130 (1988): 105.

34. Huchra, Geller, de Lapparent, and Burg *Astronomical Union Symposium Series*.

35. Calder and Lahav, *Astronomy & Geophysics* 49 (2008) 1:13–1:18.

36. The actual reason why the cosmic web reaches larger scales when dark energy is present is slightly subtle, and related to how the balance between matter and radiation plays out differently in the early universe when dark energy is present.

37. Tulin and Yu, *Physics Reports* 730 (2018): 1.

38. Pontzen and Governato, *Nature* 506, no. 7487 (2014): 171; Pontzen and Peiris, *New Scientist* 2772 (2010): 22.

39. Abel, Bryan, and Norman, *Science* 295, 5552 (2002): 93.

CHAPTER 3: GALAXIES AND THE SUB-GRID

1. Tinsley, "Evolution of Galaxies and Its Significance for Cosmology," PhD thesis, University of Texas at Austinj (1967), http://hdl.handle.net/2152/65619.

2. Sandage, *The Observatory* 99 (1968): 91.

3. "The Supereyes: Five Giant Telescopes Now in Construction to Advance Astronomy," *Wall Street Journal*, October 10, 1967.

4. Hill, *My Daughter Beatrice* (American Physical Society, 1986), 49.

5. Tinsley, "Evolution of Galaxies."

6. Cole Catley, *Bright Star: Beatrice Hill Tinsley, Astronomer* (Cape Catley Press, 2006), 165.

7. Sandage, *The Observatory*; see also Oke and Sandage, *Astrophysical Journal* 154 (1968): 21.

8. Tinsley, *Astrophysics and Space Science* 6, no. 3 (1970): 344.

9. Sandage, *Astrophysical Journal* 178 (1972): 1.

10. Bartelmann, *Classical and Quantum Gravity* 27, 233001 (2010).

11. Peebles, *Astrophysical Journal Letters* 263 (1982): L1; Blumenthal et al., *Nature* 311 (1984): 517; Frenk et al., *Nature* 317 (1985): 595.

12. White, in "The Epoch of Galaxy Formation," *NATO Advanced Science Institutes (ASI) Series C* 264 (1989): 15.

13. White and Frenk, *Astrophysical Journal*, 379 (1991): 52.

14. Sanders, "Mass Discrepancies in Galaxies: Dark Matter and Alternatives," *Astronomy and Astrophysics Review* 2 (1990): 1.

15. See, for example, the discussions in White, "Epoch of Galaxy Formation."

16. Ellis, in "The Hubble Deep Field," *STScI Symposium Series 11* (Cambridge University Press, 1988), 27.

17. Adorf, "The Hubble Deep Field Project," *ST-ECF Newsletter* 23 (1995): 24.

18. Larson, *Monthly Notices of the Royal Astronomical Society* 169 (1974): 229; Larson and Tinsley, *Astrophysical Journal* 219 (1977): 46.

19. Somerville, Primack, and Faber, *Monthly Notices of the Royal Astronomical Society* 320 (2001): 504.

20. Ellis, "The Hubble Deep Field."

21. Tinsley, *Fundamentals of Cosmic Physics* 5 (1980): 287.

22. Cen, Jameson, Liu, and Ostriker, *Astrophysical Journal* 362 (1990): L41.

23. Gingold and Monaghan, *Monthly Notices of the Royal Astronomical Society* 181 (1977): 375.

24. Monaghan, *Annual Reviews in Astronomy and Astrophysics* 30 (1992): 543.

25. Monaghan, Bicknell, and Humble, *Physical Review D* 77 (1994): 217.

26. Katz and Gunn, *Astrophysical Journal* 377 (1991): 365; Navarro and Benz, *Astrophysical Journal* 380 (1991): 320.

27. Katz, *Astrophysical Journal* 391 (1992): 502.

28. Moore et al., *Astrophysical Journal* 524, no. 1 (1999): L19.

29. Ostriker and Steinhardt, *Science* 300, no. 5627 (2003): 1909.

30. Battersby, *New Scientist* 184, no. 2469 (2004): 20.

31. Governato et al., *Astrophysical Journal* 607 (2004): 688; Governato et al., *Monthly Notices of the Royal Astronomical Society* 374 (2007): 1479.

32. Katz, *Astrophysical Journal*.

33. Springel and Hernquist, *Monthly Notices of the Royal Astronomical Society* 339 (2003): 289; Robertson et al., *Astrophysical Journal* 645 (2006): 986.

34. Stinson et al., *Monthly Notices of the Royal Astronomical Society* 373, no. 3 (2006): 1074.

35. Governato et al., *Monthly Notices of the Royal Astronomical Society* 374 (2007): 1479.

36. Pontzen and Governato, *Monthly Notices of the Royal Astronomical Society* 421 (2012): 3464.
37. Kauffmann, *Monthly Notices of the Royal Astronomical Society* 441 (2014): 2717.

CHAPTER 4: BLACK HOLES

1. Interview of Martin Schwarzschild by Spencer Weart on March 10, 1977, Niels Bohr Library and Archives, American Institute of Physics, aip.org/history-programs /niels-bohr-library/oral-histories/4870-1.
2. In fact, Schwarzschild based the first of these papers on an incomplete, previously published version of Einstein's equations, so his swift application of a brand-new idea was not quite as extraordinary as it first appears.
3. Schwarzschild, *Sitzungsberichte der Königlich Preussischen Akademie der Wissenschaften zu Berlin, Phys. Math. Klasse* (1916): 424.
4. Schwarzschild, *Gesammelte Werke Collected Works* (Springer, 1992).
5. Thorne, *From Black Holes to Time Warps: Einstein's Outrageous Legacy* (W. W. Norton, 1994).
6. Oppenheimer and Snyder, *Physical Review* 56 (1939): 455.
7. Oppenheimer and Volkoff, *Physical Review* 55 (1939): 374.
8. Bird and Sherwin, *American Prometheus: The Triumph and Tragedy of J. Robert Oppenheimer* (Alfred A. Knopf, 2005).
9. Arnett, Baym, and Cooper, "Stirling Colgate," *Biographical Memoirs of the National Academy of Sciences* (2020).
10. Arnett, Baym, and Cooper, "Stirling Colgate."
11. Teller, *Memoirs: A Twentieth-Century Journey in Science and Politics* (Perseus Publications, 2001), 166.
12. Arnett, Baym, and Cooper, "Stirling Colgate."
13. Colgate, *Canadian Journal of Physics* 46, no. 10 (1968): S476; Klebesadel, Strong, and Olson, *Astrophysical Journal* 182 (1973): L85.
14. Breen and McCarthy, *Vistas in Astronomy*, 39 (1995): 363.
15. May and White, *Physical Review Letters* 141 (1996): 4.
16. Hafele and Keating, *Science* 177 (1972): 168.
17. Han Fei, *The Complete Works of Han Fei Tzu* (Arthur Probsthain, II, ca. 300 BCE), 204.
18. Einstein and Rosen, *Physical Review* 49 (1935): 404.
19. Hannam et al., *Physical Review D* 78, 064020 (2008).
20. Thorne (2017), Nobel Lecture, NobelPrize.org, Nobel Prize Outreach AB 2023, nobelprize.org/prizes/physics/2017/thorne/lecture, accessed October 28, 2022.
21. Wheeler, *Physical Review* 97 (1955): 511.
22. Murphy, *Women Becoming Mathematicians* (MIT Press, 2000).
23. Hahn, *Communications on Pure and Applied Mathematics* 11, no. 2 (1958): 243.

24. Lindquist, "The Two-Body Problem in Geometrodynamics," PhD thesis, Princeton University (1962): 24.

25. Hahn and Lindquist, *Annals of Physics* 29 (1964): 304.

26. Pretorious, *Physical Review Letters* 95 (2005): 121101; Campanelli et al., *Physical Review Letters* 96 (2006): 111101; Baker et al., *Physical Review Letters* 96 (2006): 111102.

27. Overbye, *Lonely Hearts of the Cosmos* (HarperCollins, 1991); Schmidt, *Nature* 197, no. 4872 (1963): 1040; Greenstein and Thomas, *Astronomical Journal* 68 (1963): 279.

28. Blandford and Znajek, *Monthly Notices of the Royal Astronomical Society* 179 (1977): 433.

29. Springel and Hernquist, *Monthly Notices of the Royal Astronomical Society* 339, no. 2 (2003): 289.

30. Di Matteo, private communication (2020).

31. Di Matteo, Springel, and Hernquist, *Nature* 433 (2005).

32. Di Matteo, private communication (2020).

33. Silk and Rees, *Astronomy & Astrophysics* 331 (1998): L1.

34. Magorrian et al., *Astronomical Journal* 115 (1998): 2285.

35. Sanchez et al., *Astrophysical Journal* 911 (2021): 116; Davies et al., *Monthly Notices of the Royal Astronomical Society* 501 (2021): 236.

36. Volonteri, *Astronomy and Astrophysics Review* 18 (2010): 279.

37. Tremmel et al., *Astrophysical Journal* 857 (2018): 22.

38. ESA (2021), *LISA Mission Summary*, sci.esa.int/web/lisa/-/61367-mission-summary, accessed October 29, 2022.

39. Bell Burnell, *Science* 304, no. 5670 (2004): 489

40. Hawking, "Properties of Expanding Universes," PhD thesis, University of Cambridge (1966), doi.org/10.17863/CAM.11283.

CHAPTER 5: QUANTUM MECHANICS AND COSMIC ORIGINS

1. Louis de Broglie—"Biographical," NobelPrize.org, Nobel Prize Outreach AB 2023, nobelprize.org/prizes/physics/1929/broglie/biographical, accessed October 28, 2022.

2. Islam et al., *Chemical Society Reviews* 43 (2014): 185; Csermely et al., "Structure and Dynamics of Molecular Networks: a Novel Paradigm of Drug Discovery," *Pharmacology & Therapeutics* 138 (2013): 333; Gur et al., *Journal of Chemical Physics* 143, 075101; Qu et al., *Advances in Civil Engineering* 1687 (2018); Hou et al., *Carbon* 115 (2017): 188.

3. Hubbard, in *Discovering Reality*, eds. Harding and Hintikka (Schenkman Publishing Co., 1979), 45–69.

4. Marquard, "Ruth Hubbard, 92, First Woman Tenured in Biology at Harvard," *Boston Globe*, September 4, 2016, bostonglobe.com/metro/2016/09/04/ruth-hubbard-first-woman-tenured-biology-harvard/zdiRSECiShE4rAJaCBNNTP/story.html.

5. Karplus, *Annual Reviews in Biophysics and Biomolecular Structure* 35 (2006): 1.

6. Karplus, *Biophysics and Biomolecular Structure*.

7. In reality, quantum simulations represent variations through space in a much more complex, carefully crafted way, but imagining a grid is good enough for understanding the overall approach.

8. Miller, *Physics Today* 66, no. 12 (2013): 13.

9. Benioff, *International Journal of Theoretical Physics* 21, no. 3 (1982): 177.

10. Feynman, *International Journal of Theoretical Physics* 21, no. 6 (1982): 467.

11. *Restructure!* blog (August 7, 2009), restructure.wordpress.com/2009/08/07/sexist -feynman-called-a-woman-worse-than-a-whore, accessed October 28, 2022.

12. Lloyd, *Science* 273 (1996): 1073.

13. Google AI Quantum et al., *Science* 369, no. 6507 (2020): 1084.

14. Preskill, "Quantum Computing in the NISQ Era and Beyond," *Quantum* 2 (2018): 79.

15. Heuck, Jacobs, and Englund, *Physical Review Letters* 124, 160501 (2020).

16. Byrne, *The Many Worlds of Hugh Everett III* (Oxford University Press, 2010).

17. Matteucci et al., *European Journal of Physics* 34 (2013): 511.

18. Von Neumann, *Mathematical Foundations of Quantum Mechanics: New Edition*, ed. Nicholas A. Wheeler (Princeton University Press, 2018), 273.

19. For a summary of some of the remarkable experimental work that shows these principles in action, see Zeilinger, *Review of Modern Physics* 71 (1999): S288.

20. Wigner, in *The Collected Works of Eugene Paul Wigner* B, no. 6: 261 (Springer, 1972).

21. For a discussion of various shades of idealism, see Guyer and Horstmann, "Idealism," in *The Stanford Encyclopedia of Philosophy*, ed. Edward N. Zalta (2022), retrieved from plato.stanford.edu/archives/spr2022/entries/idealism.

22. Wheeler and Zurek, in *Quantum Theory and Measurement* (Princeton University Press, 1983), 182, jstor.org/stable/j.ctt7ztxn5.

23. Penrose, *The Emperor's New Mind* (Oxford University Press, 1989).

24. Howl, Penrose, and Fuentes, *New Journal of Physics* 21, no. 4 (2019): 043047.

25. Aspect, Dalibard, Roger, *Physical Review Letters* 49, no. 25 (1982): 1804.

26. The full thesis was not published until 1973; Everett, in *The Many-Worlds Interpretation of Quantum Mechanics* (Princeton University Press, 1973), 3.

27. Saunders, *Foundations of Physics* 23, no. 12 (1993): 1553.

28. Deutsch, *Proceedings of the Royal Society* A 400, no. 1818 (1985): 97.

29. For an extensive discussion from both pro- and anti-Everettians, see Saunders, Barrett, Kent, and Wallace, *Many Worlds?* (Oxford University Press, 2010).

30. See, for example, chapter 27 of Penrose, "The Road to Reality," Jonathan Cape (2004).

31. US Bureau of Labor Statistics CPI Inflation Calculator, June 1994 to June 2022, bls.gov/data/inflation_calculator.htm, retrieved December 3, 2022.

32. Bresciani-Turroni, *The Economics of Inflation* (Bradford & Dickens, 1937), 441.

33. Calculating this minimum requirement involves comparing the size of the observable universe today to the way that light beams travel across the young universe.

34. For an up-to-date assessment see Planck Collaboration, *Astronomy & Astrophysics* 641 (2018) A6.

35. One might hope that the layout of structure seen in the cosmic microwave background, which shows us the specific ripples in our particular universe, could be used as the unique "correct" starting point for a simulation. However, because the light is so ancient, it has also traveled a very long distance. It tells us about the starting point only for very distant parts of the universe, and we do not know how the galaxies in those regions turned out.

36. Springel et al., *Monthly Notices of the Royal Astronomical Society* 475 (2018): 676; Tremmel et al., *Monthly Notices of the Royal Astronomical Society* 470 (2018): 1121; Schaye et al., *Monthly Notices of the Royal Astronomical Society* 446 (2015): 521.

37. Roth, Pontzen, and Peiris, *Monthly Notices of the Royal Astronomical Society* 455 (2016): 974.

38. Rey et al., *Astrophysical Journal* 886, no. 1 (2019): L3; Pontzen et al., *Monthly Notices of the Royal Astronomical Society* 465 (2017): 547; Sanchez et al., *Astrophysical Journal* 911, no. 2 (2021): 116.

39. Pontzen, Slosar, Roth, and Peiris, *Physical Review D* 93 (2016): 3519.

40. Angulo and Pontzen, *Monthly Notices of the Royal Astronomical Society* 462, no. 1 (2016): L1.

41. Mack, *The End of Everything* (Allen Lane, 2020).

42. This argument has been made multiple times in different guises; see, for example, chapter 28:5 of Penrose, "The Road to Reality." See also Ijjas et al., "Pop Goes the Universe," *Scientific American* 316, no. 32 (2017).

43. Kamionkowski and Kovetz, *Annual Review of Astronomy and Astrophysics* 54 (2016): 227.

44. Giddings and Mangano, *Physical Review D* 78, no. 3, 035009 (2008); Hut and Rees, *Nature* 302, no. 5908 (1983): 508.

CHAPTER 6: THINKING

1. Homer, *Odyssey* (c. eighth century BCE), 7, 87.

2. Lapin, "NYPD's Robot Dog Will Be Returned after Outrage," *New York Post*, April 28, 2021.

3. *Guardian News* online (2018), "Human v Robot Dog: Boston Dynamics Takes on Its Door-Opening SpotMini," youtube.com/watch?v=W1LWMk7JB80, accessed October 28, 2022.

4. Daniel C. Dennett's book *Consciousness Explained* (1991) and Douglas Hofstadter's *Gödel, Escher, Bach* (1979) suggest that consciousness may anyway be a natural consequence of sophisticated thinking apparatus.

5. Turing, *Mind* 49 (1950): 433.

6. The Law Society (December 13, 2018), "Six Ways the Legal Sector Is Using AI Right

Now," lawsociety.org.uk/campaigns/lawtech/features/six-ways-the-legal-sector-is
-using-ai, accessed February 3, 2022.

7. Assuming a 250-megabit-per-second digital cinema package format, and ninety-minute films. That's perfectly long enough.

8. National Library of Medicines Profiles in Science: Joshua Lederberg biographical overview, profiles.nlm.nih.gov/spotlight/bb/feature/biographical-overview, accessed October 28, 2022.

9. Blumberg, *Nature* 452 (2008): 422.

10. More precisely, electric and magnetic fields are used to infer the mass-to-charge ratio of the fragments.

11. Bielow et al., *Journal of Proteome Research* 10, no. 7 (2011): 2922.

12. Waddell Smith, in *Encyclopedia of Forensic Sciences* (Academic Press, 2013): 603.

13. The mission was due to be launched on a Russian rocket in 2022 but was suspended due to the war with Ukraine. It is now expected to launch late in the decade. "Rover Ready—Next Steps for ExoMars," esa.int/Science_Exploration/Human_and _Robotic_Exploration/Exploration/ExoMars/Rover_ready_next_steps_for _ExoMars, accessed October 28, 2022.

14. Planck Collaboration, *Astronomy and Astrophysics Review* 641 (2020): 6.

15. Joyce, Lombriser, Schmidt, *Annual Review of Nuclear and Particle Science* 66: 95.

16. Jaynes, *Probability Theory: The Logic of Science* (Cambridge University Press, 2003), 112.

17. There are literally hundreds of papers applying these techniques. For some foundational examples, see: Ashton et al. , *Astrophysical Journal Supplement* 241 (2019): 27; Verde et al., *Astrophysical Journal Supplement* 148 (2003): 195; Kafle, *Astrophysical Journal* 794 (2014): 59.

18. Lightman and Brawer, *The Lives and Worlds of Modern Cosmologists* (Harvard University Press, 1992).

19. Hawking, "On the Rotation of the Universe," *Monthly Notices of the Royal Astronomical Society* 142, no. 2 (1969): 129.

20. Hawking, "Rotation of the Universe."

21. Pontzen, *Physical Review D* 79, no. 10 (2009): 103518; Pontzen and Challinor, *Monthly Notices of the Royal Astronomical Society* 380 (2007): 1387.

22. Hayeden and Villeneuve, *Cambridge Archeological Journal* 21, no. 3 (2011): 331.

23. Coe et al., *Astrophysical Journal* 132 (2006): 926.

24. Fan and Makram, *Frontiers in Neuroinformatics* 13 (2019): 32.

25. Hodgkin and Huxley, *Journal of Physiology* 117 (1952): 500.

26. Swanson and Lichtman, *Annual Reviews of Neuroscience* 39 (2016): 197.

27. As previously stated, the 2021 capacity was around 8×10^{21} bytes. See *Statista*, statista.com.

28. Hebb, *The Organization of Behavior: A Neurophysical Theory* (Wiley & Sons, 1949); Martin, Grimwood, and Morris, *Annual Reviews of Neuroscience* 23 (2000): 649.

29. Hebb, *Journal of General Psychology* 21, no. 1 (1939): 73.

30. Fields, *Scientific American* 322 (2020): 74.

31. Bargmann and Marder, *Nature Methods* 10 (2013): 483; Jabr, "The Connectome Debate: Is Mapping the Mind of a Worm Worth It?" *Scientific American* (October 2, 2012).

32. Rosenblatt, *Research Trends of Cornell Aeronautical Laboratory* VI (1958): 2.

33. "Electronic 'Brain' Teaches Itself," *New York Times*, July 13, 1958.

34. Rosenblatt, *Principles of Neurodynamics: Perceptrons and the Theory of Brain Mechanisms*, Cornell Aeronautical Laboratory Report VG-1196-G-8 (1961).

35. Lefkowitz, "Professor's Perceptron Paved the Way for AI—60 Years Too Soon," *Cornell Chronicle*, September 25, 2019, news.cornell.edu/stories/2019/09/professors -perceptron-paved-way-ai-60-years-too-soon, accessed October 28, 2022.

36. Cornell University News Service records, #4-3-15, Mark I Perceptron at Cornell Aeronautical Laboratory, accessed February 22, 2023, Photograph, Cornell University Library, Division of Rare and Manuscript Collections, digital.library.cornell .edu/catalog/ss:550351.

37. Hay, *Mark 1 Perceptron Operators' Manual*, Cornell Aeronautical Laboratory Report VG-1196-G-5.

38. Crawford, *Atlas of AI* (Yale University Press, 2021).

39. Firth, Lahav, and Somerville, *Monthly Notices of the Royal Astronomical Society* 339 (2003): 1195; Collister and Lahav, *Publications of the Astronomical Society of the Pacific* 116 (2004) 345.

40. de Jong et al., *Astronomy and Astrophysics* 604 (2017): A134.

41. Lochner et al., *Astrophysical Journal Supplement* 225 (2016): 31.

42. Schanche et al., *Monthly Notices of the Royal Astronomical Society* 483, no. 4 (2019): 5534.

43. Jumper et al., *Nature* 596 (2012): 583.

44. Anderson, "The End of Theory: The Data Deluge Makes the Scientific Method Obsolete," July 16, 2008, *Wired*, wired.com/2008/06/pb-theory, accessed October 28, 2022.

45. Matson, "Faster-Than-Light Neutrinos? Physics Luminaries Voice Doubts," September 26, 2011, *Scientific American*.

46. Reich, "Embattled Neutrino Project Leaders Step Down," *Nature* (2012), doi: 10.1038/nature.2012:10371.

47. GDPR article 15 1(h), "Right of Access by the Data Subject," gdpr.eu/article-15 -right-of-access, accessed October 28, 2022.

48. Iten et al., *Physical Review Letters* 124, 010508 (2020).

49. Ruehle, *Physics Reports* 839 (2019): 1.

50. Lucie-Smith et al., *Physical Review D* 105, no. 10 (2022): 103533.

51. Cellan-Jones, "Robots 'to Replace up to 20 Million Factory Jobs' by 2030," June 26, 2019, *BBC News*, bbc.co.uk/news/business-48760799, accessed October 28, 2022.

52. Buolamwini, "Artificial Intelligence Has a Problem with Gender and Racial Bias. Here's How to Solve It," February 7, 2019, *Time*, time.com/5520558/artificial-intelligence-racial-gender-bias, accessed October 28, 2022.

53. Crawford, *Atlas of AI*.

54. Swain, "Twitter Admits Far More Russian Bots Posted on Election Than It Had Disclosed," *Guardian*, January 20, 2018, theguardian.com/technology/2018/jan/19/twitter-admits-far-more-russian-bots-posted-on-election-than-it-had-disclosed, accessed October 28, 2022.

55. Brown et al., "Language Models Are Few-Shot Learners," in *Advances in Neural Information Processing Systems* 33 (2020), https://arxiv.org/abs/2005:14165v1.

56. Floridi and Chiriatti, *Minds & Machines* 30 (2020): 681; for an impressive example of a system based on GPT that can write computer code, see GitHub Copilot, github.com/features/copilot.

CHAPTER 7: SIMULATIONS, SCIENCE, AND REALITY

1. Fredkin, *International Journal of Theoretical Physics* 42 (2003): 2; Lloyd (2005), *Programming the Universe*, Jonathan Cape.

2. Greene, "Is Our Universe a Simulation?" November 4, 2016, startalkmedia.com/show/universe-simulation-brian-greene, *StarTalk Radio*; Powell, "Elon Musk Says We Live in a Simulation. Here's How We Might Tell If He Is Right," October 2, 2018, *NBC News*, nbcnews.com/mach/science/what-simulation-hypothesis-why-some-think-life-simulated-reality-ncna913926; Dawkins, "Are Our Heads in the Cloud? Science Fiction or Fact?" Richard Dawkins website, richarddawkins.com/articles/article/are-our-heads-in-the-cloud.

3. Bostrom, *Philosophical Quarterly* 53 (2003): 243.

4. Chalmers, *Reality+* (Allen Lane, 2021).

5. Dawkins, "Are Our Heads in the Cloud?"

6. Since these were human computations, the idea of a bit is not entirely natural, but one can still make an estimate. Holmberg had seventy-four light bulbs, each free to move in two dimensions and with two dimensions of motion. That makes 296 numbers to describe the simulation at any given moment. Assuming he was able to measure to three significant figures, the number of bits per number is approximately $\log_2 103 \approx 10$. Roughly, this gives 3,000 bits in total. As for the Richardsons, the initial grid for which Louis Fry gives explicit credit to his wife has seventy values for wind recorded to three significant figures, and forty-five values for pressure to four significant figures (as well as other ancillary information). This gives an estimate of 1,000 bits.

7. Raju, *Physics Reports* 943 (2022): 1.

8. Hawking, *Physical Review Letters* 13 (1976): 191; Bekenstein, *Physics Today* 33, no. 1 (1980): 24; Zurek and Thorne, *Physical Review Letters* 54, no. 20 (1985): 2171. Seth Lloyd in his book quotes a much smaller number of 10^{92} qubits, based on

calculating the entropy of a thermal state in the absence of any gravity. Whichever way one chooses to make the calculation, the basic conclusion—that only the whole universe can simulate the whole universe—is unaffected.

9. Preskill, "Quantum Computing."

10. The idea can be traced back to Wheeler, *Quantum Coherence and Reality* (World Scientific, 1992), 281.

11. Barrow, *Universe or Multiverse?* (Cambridge University Press, 2007), 481; Beane et al., *European Physics Journal A* 50 (2014): 148.

12. Albert Einstein, *Königlich Preußische Akademie der Wissenschaften, Sitzungsberichte* (1915), 831.

13. Dyson, Eddington, and Davidson, *Philosophical Transactions of the Royal Society of London Series A* 220 (1920): 291.

14. Margaret Morrison, *Philosophical Studies* 143 (2009): 33; see also Norton and Suppe, in *Changing the Atmosphere: Expert Knowledge and Environmental Governance* (MIT Press, 2001).

15. Pontzen et al., *Monthly Notices of the Royal Astronomical Society* 465 (2017): 547.

INDEX